The Sassafras Science Adventures

Volume 3: Botany

Johnny Congo & Paige Hudson

THE SASSAFRAS SCIENCE ADVENTURES
VOLUME 3: BOTANY

Second Printing 2015
(First Printing 2014)
Copyright @ Elemental Science, Inc.
Email: info@elementalscience.com

ISBN: 978-1-935614-33-3
Cover Design by Paige Hudson & Eunike Nugroho
Illustrations by Eunike Nugroho (be.net/inikeke)

Printed In USA For World Wide Distribution

No part of this book may be reproduced or transmitted in any form or by any means, electronic or mechanical, including photocopying, recording, or by means of any information storage and retrieval system, without permission in writing from the authors. The only exception is brief quotations in printed reviews.

For more copies write to :
Elemental Science
610 N. Main St #207
Blacksburg, VA 24060
info@elementalscience.com

Dedication

We dedicate this book to all the students who have fallen in love with the Sassafras twins. Thank you for continuing your adventure with Blaine and Tracey!

Volume 3: Botany
Table of Contents

Chapter 1: The Basics of Botany — 1
Memories on the Horse Swing.............1
Next Up—The Study of Plants................9

Chapter 2: Return to the Jungle — 17
Falling Orchids............17
Fishing for Ferns............25

Chapter 3: Kidnapped — 35
Lilies on the Trail............35
Fighting Fungi............44

Chapter 4: Lost in a Scottish Castle Maze — 55
Roses for the Lady............55
A Maze of Boxwoods............68

Chapter 5: The Mystery of the Stolen Roses — 77
Mossy Mysteries............77
Peat-filled Pathways............88

Chapter 6: The Gaucho with Ten Names — 97
Argentinian Pampas............97
Games and Grasses............107

Chapter 7: Who Killed the Gray Fox? — 117
Wrangling Wildflowers............117
Ombú Occupation............128

The Sassafras Science Adventures

Chapter 8: Pirates in Borneo! 137
Picture-perfect Palms..........................137
Carnivorous Captors.............................146

Chapter 9: Escape into the Jungle 157
Marauding Molds..................................157
Running toward the Giant Rafflesia..166

Chapter 10: The Secret Siberian Railway 175
Shrubs and Stowaways.......................175
Capturing the Crocus...........................183

Chapter 11: The End of the Line 193
Look-out for Lichens............................193
Accidental Algae..................................201

Chapter 12: The Mysterious Miss Été 209
Plant Cells in Paris...............................209
The Photosyntactic Phantom Phone...220

Chapter 13: A Tour of Versailles 229
Crashing Chestnuts.............................229
The Apple Festival...............................238

Chapter 14: California Creatures of Mystery 247
Squatchin' with the Sitkas..................247
Cones of a Different Origin................257

Chapter 15: Sasquatch Sighting! 267
Exploring the Redwoods.....................267
Magic Vanishing Mushrooms............276

The Sassafras Science Adventures

Chapter 16: The Glitch Is Gone — 287

Joshua Tree National Park..............287
Evidence of the Legendary Cactus Head......................................297

Chapter 17: Solving the Cactus Head Mystery — 305

Creeping Creosote...............................305
Protruding Paddle Cacti...................314

Chapter 18: Back to the Basement — 323

Surprise Jungle....................................323
Bonus Data...328

Make the most of your journey with the Sassafras Twins!

Add our activity guide, logbook, or lapbooking guide to create a full science curriculum for your students!

The Sassafras Guide to Botany includes chapter summaries and an array of options that coordinate with the individual chapters of this novel. This guide provides ideas for experiments, notebooking, vocabulary, memory work, and additional activities to enhance what your students are learning about plants!

The Official Sassafras SCIDAT Logbook: Botany Edition partners with the activity guide to help your students document their journey throughout this novel. The logbook includes their own SCIDAT log pages as well as biome fact sheets and a botany glossary.

Lapbooking through Botany with the Sassafras Twins provides a gentle option for enhancing what your students are learning about plants through this novel. The guide contains a reading plan, templates, and pictures to create a beautiful lapbook on botany, vocabulary, and coordinated scientific demonstrations!

Visit SassafrasScience.com to learn more!

The Sassafras Guide to the Characters

Throughout the Book*

- ★ **Blaine Sassafras** – The male Sassafras twin, also known as Train. He started the summer hating science, but thanks to the zip lines, he is enjoying experiencing science face-to-face.
- ★ **Tracey Sassafras** – The female Sassafras twin, also known as Blaisey. She started the summer hating science, but thanks to the zip lines, she is also enjoying experiencing science face-to-face.
- ★ **Uncle Cecil** – The Sassafras twins' crazy, forgetful, and messy uncle. He is the scientist behind the invisible zip lines.
- ★ **President Lincoln** – Uncle Cecil's lab assistant, who also happens to be a prairie dog. He is also the co-inventor of the zip lines.
- ★ **The Man with No Eyebrows** – He has no eyebrows and an extreme dislike of Uncle Cecil. Not only is he spying on the red-haired scientist, but he is also trying to sabotage the twins at every stop.

(*__Note__ – These characters also appeared in *The Sassafras Science Adventures Volume 1: Zoology* and *The Sassafras Science Adventures Volume 2: Anatomy*.)

Cecil's Neighborhood (Chapter 1)

- ★ **Mrs. Pascapli (paz-kah-pah-LEE)** – She lives at 1106 North Pecan Street, next door to Uncle Cecil.

Peru (Chapters 2 & 3)

- ★ **Arrio** – The native Peruvian who serves as the local expert for the twins in the Amazon Rainforest. He is a full-fledged Yora tribesman and friend of Alvaro. He also appeared in *The Sassafras Science Adventures Volume 1: Zoology*.
- ★ **Itotia** – The leader of the Matsigenka tribe, who are the

enemies of the Yora.
* **Tenyoa** – A tracker with the Yora tribe.
* **Alvaro Manihuari** – The owner of the Out-on-a-Limb guesthouse. He is a friend of Arrio and the Yora tribesman. He also appeared in *The Sassafras Science Adventures Volume 1: Zoology*.
* **Ernesto Perez** – The president of the ProLog operations in Peru. He also appeared in *The Sassafras Science Adventures Volume 1: Zoology*.

SCOTLAND (CHAPTERS 4 & 5)
* **Fiona McRay** – The resident botanist at Dockerty Castle. She is also the twins' local expert as they explore the Scottish castle's gardens and nearby peat bog.
* **Dunmore** – The butler at Dockerty Castle.
* **Osla** – The maid at Dockerty Castle.
* **Rona** – The baker at Dockerty Castle.
* **Angus** – The gardener at Dockerty Castle.
* **Lief** – The piper at Dockerty Castle.
* **Lady Dockerty** – The mistress of Dockerty Castle.
* **Sir Kentalot** – Lady Dockerty's brother.
* **Miles Dockerty** – Lady Dockerty's son and judge of the Take Our Breath Away singing competition. He also appeared in *The Sassafras Science Adventures Volume 2: Anatomy*.

THE SASSAFRAS SCIENCE ADVENTURES

Argentina (Chapters 6 & 7)

- **Felipe Moreno** – The entertainer and pianist at the Cantina de Pampas. He is the local expert as the twins explore the Argentinian pampas.
- **Emilio** – The bartender at the Cantina de Pampas.
- **Raul Juan Pablo Eduardo Santiago Mateo De La Casillas . . . the third** – The gaucho with ten names who is on a quest to discover who killed his grey fox.
- **Darts Domingo** – The rough darts player who challenges the gaucho with ten names on his quest.
- **Jorge Alfonzo** – A trapper on the pampas who confronts the gaucho with ten names on his quest.
- **Franco Lorenzo** – An Argentinian cattle wrangler who tests the gaucho with ten names on his quest.
- **Manuel Hernandez** – The owner of the Hacienda de Hernandez, a white mansion in the middle of the Argentinian pampas.
- **Nicolette Hernandez** – The wife of Manuel Hernandez.

Borneo (Chapters 8 & 9)

- **Trisno Kanang** – The twins' local expert in Borneo. He works at Pitchers Beachside Resort.
- **Novi Anita** – The hotel manager at Pitchers Beachside Resort.
- **Rover and Zaza Ridgeburn** – A couple who have come to celebrate their fiftieth anniversary at the Pitchers Beachside Resort.
- **Rama** – The leader of the pirates who attack the Pitchers Beachside Resort.

Siberia (Chapters 10 & 11)

- **Pavel Markoff** – The engineer of the secret Siberian Railway train. He is the twins' local expert as they journey through the Siberian tundra.
- **Yuri Checkoff** – The conductor of the secret Siberian Railway train.

- ★ **Yuroslav Bogdanovich** – The Aggrandizer inventor and crazed Siberian scientist.
- ★ **Sveta Corvette** – The neon green punk-rocker who travels the trains as a stowaway.

France (Chapters 12 & 13)

- ★ **Été Plage** – The mysterious local expert during the twins' visit to France. They have met her multiple times in their adventures, but they usually find her in a much colder place.

Northern California (Chapters 14 & 15)

- ★ **Brock Hoverbreck** – The twins' local expert as they explore the redwoods of Northern California. He is a park ranger and expert of the flora there.
- ★ **Melody Albermully** – The leader of the C.O.M. Crew.
- ★ **Harmony Albermully** – Melody's sister and the C.O.M. Crew historian.
- ★ **Rip** – The tracker for the C.O.M. Crew.
- ★ **Sam** – The tech specialist for the C.O.M. Crew.
- ★ **Chorus (a.k.a. Cory) Albermully** – The technical assistant to Sam, and brother of Melody and Harmony.
- ★ **Ned** – The driver for the C.O.M. Crew.

Southern California (Chapters 16 & 17)

- ★ **Symphony Douglas** – The twins' local expert as they explore the Mojave Desert of Southern California. She is a park ranger at Joshua Tree National Park and cousin of Brock Hoverbreck.

Volume 3:

Botany

Chapter 1: The Basics of Botany

Memories on the Horse Swing

She smiled as the wind whipped gently through her hair. Her upward motion gracefully stalled before back down she went, in the opposite direction. A peaceful smile had found its way to her face, and there was no sign of it leaving any time soon. Tracey Sassafras gripped the rope tightly, happily kicked out her legs, and continued her relaxing ride on the horse-shaped tire swing.

Her uncle and brother had gone back inside, so she was left alone in the backyard with nothing to do but swing and daydream. She and her twelve-year-old twin, Blaine, had arrived at their Uncle Cecil's doorstep only a couple of weeks before, though it seemed like much longer than that now. They had both failed their science class in school this past year, so their concerned parents had packed them up on a bus and sent them to their uncle's home. The plan was to make Blaine and Tracey spend the duration of their summer break studying science again.

As they stepped off the bus, the twins had resigned themselves to spending the time being bored to tears. They knew their uncle was a bit of an eccentric scientist, but they were sure that even he couldn't make science fun. The twins were shocked to discover that Uncle Cecil had invented invisible zip lines that had the ability to whisk a person to any location on the globe at the speed of light. It was Uncle Cecil's belief that the best way for the kids to learn science was not sitting in a classroom with their heads in a book. Instead, they should be out zipping around the globe experiencing science face-to-face. That is how he believed they would not only learn but also fall in love with science.

All one needed to make these lines work was a specially designed three-ringed carabiner. One ring was set to the desired longitude coordinate, one ring was set to the latitude coordinate, and the third ring locked the carabiner. With this device, the person was also kept safely connected in the harness to the invisible zip line. When Cecil had first explained all of this to the twins, they had thought their uncle had truly lost his mind. However, they had humored him, put on the harnesses, calibrated their three-ringed carabiners, and zipped across the invisible lines themselves. As irrational as it sounded, it really did work. The twins had spent the past two weeks using the unseen lines to zip to different locations all over the planet studying scientific topics related to zoology and anatomy.

At the start of their zip-lining adventures, they had despised science with a passion. The twins had considered it to be a long list of boring facts, but as they were encountering science face-to-face, their disposition toward it was changing. These days, Blaine and Tracey not only liked science but were also learning to love it.

The only hindrance on their path of increased desire to learn science was a certain mysterious man who had no eyebrows. He had tried to derail their progress at nearly every stop.   H e had left them marooned out in the grasslands of Kenya among deadly predators. He had trapped them in a cobra-infested tomb in Egypt. He had helped cut down a tree out from under them in Peru. He had sabotaged their uncle's computer—sending the twins zipping off on the invisible zip lines to two different countries. He had chased them using a disappearing magician's suit and cape. Plus, Tracey was pretty sure he was the one who had trapped her and Blaine inside a couple of virtually indestructible, soundproof boxes, which had almost caused them to be buried alive. In fact, Tracey had her suspicions that this man was always lurking around, causing problems that she and Blaine weren't even aware of.

Tracey shuddered on the swing as she pictured his scowling face. Her frown disappeared in a flash of joy and confidence as she remembered that the Man with No Eyebrows hadn't stopped them yet! She was bound and determined to keep moving forward to learn science, and she knew that Blaine felt the same way.

Tracey was lost in her memories when a soft thud broke her reverie. Her focus snapped back to the present, and she looked down at the green grass below to locate the source of the sound. She gasped as she saw that her smartphone had fallen out of her pocket.

"Whew," Tracey thought. "That could've been bad. This device is far too important to lose or damage now."

The agile twelve-year-old reached out and grabbed the phone as she passed by on the tire swing. She and Blaine both had one, which they used for more than just making calls, surfing the Internet, or playing games. The smartphones were the key to guiding the Sassafras twins on their journey. Uncle Cecil, with the help of his lab assistant prairie dog, had created several applications—the two most important being the LINLOC and SCIDAT applications. LINLOC stood for "Line Locations," and it gave the twins the longitude and latitude coordinates for each location that they were slated to visit. This application also listed the scientific topics they would be studying and the name of a local science expert who would help facilitate the twins' learning.

The other app, SCIDAT, stood for "Scientific Data." This is where the twins recorded all of the information they were collecting. At each location, they entered the data they had learned into their phones and then sent it back to Uncle Cecil. This step was very important, because if they didn't send in the scientific information correctly, they wouldn't be able to open up the LINLOC app and go to their next location. In order to progress, Blaine and Tracey had to first gather the correct scientific facts.

At first, the need to get it right had put a lot of pressure on the twins. Now, after successfully zipping their way through zoology and anatomy, recording data was second nature to them. Their devices were also equipped with an archive app, a microscope app, and high-resolution cameras. All of these tools enabled the twins to send images of the subjects they encountered to Uncle Cecil.

Tracey closed her eyes, sighed peacefully, and continued swinging happily back and forth in the cool shade cast by the tree. She let her mind wander, thinking less about how things worked and more about all of the adventures she and Blaine had been on in the last couple of weeks. The tire swing moved gracefully back and forth through the afternoon calm. She was imagining she was back in Italy, riding in a motorcycle side car through the beautiful Venetian countryside, when she was interrupted by a familiar voice. "Tracey, I've got a glass of lemonade here!"

She opened her eyes and looked toward the back porch of the house where she saw her twin brother, Blaine. He was holding a big glass full of ice-cold lemonade. "Could this day get any more relaxing?" Tracey thought to herself. "First my own quiet and peaceful turn on the tire swing, and now my brother is bringing me a glass of lemonade."

Tracey quickly hopped off the tire swing and skipped toward the back porch. She reached out to grab the waiting glass of lemony refreshment, but just as she did, Blaine jerked the glass away. He put it up to his own mouth and drank the icy-cold glass in one long gulp. Tracey's joy turned to disgust as she crossed her arms and stood silently in front of her brother with a scowl on her face. Blaine reached up, wiped his mouth, and then let out a satisfied sigh. A half-smile half-smirk formed on his face as he stood there looking at Tracey, obviously enjoying that he had fooled her.

"I said, 'I've got a glass of lemonade here,' not 'I've got a glass of lemonade here for you.'" Blaine grinned with a cocky squint.

Tracey reached up and flicked her brother on the ear. "That wasn't very nice, meanie."

"Ouch!" Blaine smiled as he reached up to rub his ear. "Ok, ok! I was just kidding. I poured a glass for you, too. It's inside on the kitchen table."

Tracey joined Blaine in laughing as they went inside. While she gulped down her own glass of lemonade, Blaine said, "The Prez is ready to give us his presentation."

Tracey nodded, knowing exactly what her brother meant. She finished the last few drops of lemonade and then followed Blaine down to the basement. Upon their arrival, they were greeted with a shout of elation from their uncle.

"Howdy-hootie! Hello! You two super-azing Sassa-ma-fras twins!" Cecil ran over to them with outstretched arms and crazy red hair sticking out every which way. He grabbed them by the shoulders and led them over to his computer desk excitedly.

"These two science whiz-kids have now successfully completed zoology and anatomy!" Cecil announced as if he were before a large crowd. Then, he held his hands up to his ears as if the nonexistent crowd had not cheered loudly enough. He looked at two plastic mannequins standing near the desk.

"Did you hear that, Socrates? Did you catch what I said, Aristotle? I said these two fanterriffic kids have successfully completed zoology AND anatomy!"

Socrates and Aristotle, the mannequins, remained still and

silent at the repeating of this wonderful news. The twins recalled how Socrates and Aristotle had started out as simple plastic skeletons. As the twins had proceeded through learning about anatomy, their uncle had added pieces to the skeletons until each had precisely represented a complete *Homo sapiens*. He had done this to show off all of the anatomy knowledge that the twins had acquired.

Cecil bounded over to the two mannequins and grabbed their arms. He made them move like they were clapping and made sounds like they were cheering for Blaine and Tracey. Over the past several weeks, the twins had become accustomed to their uncle's off-the-wall antics.

"And now that you two wonder-twins have successfully completed anatomy and zoology, we will move next to the photosyntastic subject of botany!"

Cecil raised both arms straight up into the air, with his fingers spread. "But first," he declared, leaving only one of his index fingers raised, "the President will give his presentation on anatomy."

The President that he was referring to was none other than his lab assistant, President Lincoln, the prairie dog extraordinaire. Lincoln was up on the desk with the computer mouse next to his paw, ready to give his presentation. Though they had seen proof after proof of his brilliance, the Sassafras twins still weren't exactly sure how a prairie dog had accomplished any of the feats for which Cecil gave him credit. For that matter, the twins did not even know how to communicate with him. However, here he was, once again impressing them with his abilities.

The Prez moved the mouse, and a picture of the prairie dog came up on the big screen on the wall behind the computer desk. This is what their uncle called the "tracking screen." It usually had an illuminated world map on it with two little green dots that represented the twins. When they moved from location to location around the globe, he used this map to monitor their progress. This screen was also where he received and read all of the scientific data they sent. He could also view the pictures they sent with the data.

Cecil read aloud the text that was printed over the image of the smiling prairie dog, "President Lincoln's ever-so-brief presentation on anatomy: A review of the systems of the human body and their functions."

The prairie dog tapped the mouse again as Cecil continued. "First, we have the integumentary system which covers and protects the body." Pictures of skin, sweat glands, hair, and fingernails came up on the screen, and the twins smiled. They had taken the photos while they were in the United Arab Emirates participating in a one-hundred mile horse race across the Arabian Desert called the Wind Tower 100.

"Next, we have the skeletal system," read Cecil. "This is the system that supports the body, protects organs, and permits movement." Now Blaine and Tracey saw pictures they had taken of the skeletal system when they were in Ethiopia looking for the Seven Monk Tomb and the lost Ark of the Covenant. Blaine nodded and smiled. That had been a good adventure and a fulfilling science learning experience.

"Then, we have the muscular system which moves the body and helps to support it," Cecil shared. This time, the twins saw pictures they had found using the archive application on their phones. They remembered that when it wasn't possible to take actual photographs, they could skim through the archive app to pick the appropriate pictures.

"And then, we have the nervous system which controls the body and allows a person to think and feel." There were more wonderful pictures from the archives on the phone.

"Fifth, we have the endocrine system which releases hormones that control many of the body's processes," Cecil continued. This time, pictures flashed up on the screen from the microscope application on the twins' smartphones. What a crazy leg that had been! They had been chased all over an underground lab by crazed robot squirrels. At least the twins had gotten the chance to see Summer Beach again. She was one of the twins' all-time favorite local experts!

"The circulatory system is next," Cecil read. "It carries materials to and from cells throughout the body." The twins were enjoying this review. It made them feel satisfied about all they had learned on their journey through anatomy. They could also tell how proud their uncle was of them, and even President Lincoln beamed with pride and admiration.

He clicked on the computer's mouse again as Cecil read for him. "The next system is the respiratory system. It delivers oxygen into the bloodstream." More archive images flashed across the screen.

"The eighth system to review is the digestive system. It breaks down food so that the body can use the nutrients inside. After that, we have the urinary system which removes waste materials from the body." Cecil paused, and Tracey shivered as she remembered what had happened to Blaine and her while they were studying the two systems. She wouldn't wish being trapped and buried alive in a box even on her worst enemy.

"Last but not least, we have the immune system!" Cecil stated jubilantly. "This is the system that defends the body against disease." Pictures from the twins' time in Bangkok, Thailand, came up on the screen.

Blaine looked at his sister as he said, "An apple a day keeps the doctor away. . ."

". . . that's what they say, anyway," Tracey finished, chuckling.

President Lincoln clicked the mouse one last time, bringing up the last picture. It was an image of him, together with Socrates and Aristotle, all smiling and holding up peace signs for the camera.

"Well, whippety whoppety whoo!" Cecil exclaimed. "That's a wrap for anatomy. Up next—botany, the study of plants. But before we talk about that, let's all take a walk together!"

Next Up—The Study of Plants

Wicked confidence coursed through his veins. He finally had a plan that would bring those Sassafras twins to ruin! By bringing those pesky twins to an end, he would crush all that his arch-enemy loved and cared about. He would in effect ruin Cecil Sassafras.

Long, long ago, Cecil had wronged him in a way that had left a deep and lasting mark. Some may say that what Cecil had done was an accident, but that is not how he saw it. It was no accident. It had happened because Cecil was absent-minded and googly-eyed. Now, because of that man's absent-mindedness, he had to live with the repercussions.

Over the years since the accident, he had let bitterness and revenge become his driving forces. After all the effort he had put into getting back at Cecil, things were finally about to pay off.

Over the past couple of weeks, he had failed miserably in so many ways at trying to stop those twins from learning. They had proven to be much more resilient than he had thought possible, but now he had the Dark Cape. More importantly though, he now knew how to use it. He had placed hidden cameras equipped with microphones all over Cecil's house. He had seen and heard the twins sit down to recap their adventures. In the process, they had told their uncle exactly how the Dark Cape worked, and he had heard every word.

The suit was originally designed by a magician named Phil Earp. It was as black as midnight, and included gloves, a masked helmet, and a huge cape. Phil's gimmick was to use the suit to make things disappear. He would start with small items and work his way up to bigger items, but his grandest trick was when he made himself disappear. Phil had done this by attaching something he called a "vanish string" to the inside of the cape. Simply give it a tug, and voilà, you disappeared. To reappear, you just pull the string again.

He had stolen the Dark Cape from Phil once before while in Sydney, Australia. He had tried to use it against the twins, but at that time, he had not known about the vanish string. He had made a real mess of his sabotage attempt, but that was then, and this was now.

He tightened the muscles of his hairless brow and grinned with menace. He looked at his computer monitor and tapped it to illuminate images of rooms in Cecil's house at 1104 North Pecan Street. His adversary's dwelling was a mere two doors down from his own place. He was glad to see that his hidden cameras were still working. He glanced over his shoulder and saw his harness and three-ringed carabiner lying on the floor. Wherever those twins zipped off to next, he would be there.

He stood up from his seat and slowly slipped on the magic suit. He pulled the gloves on tight, fastened the masked helmet down securely, grabbed the long flowing cape, and pulled it dramatically up around himself. Then, at the top of his lungs, shouting to no one in particular, he exclaimed, "I am the Dark

Cape!"

The Sassafras twins stood on the front porch of their uncle's house and looked at dried-up and dead potted plants. "I really should've remembered to water those," Cecil mused as he scratched his head.

The twins were sure to stand clear of the trap door they knew was in the floor of the front porch, as they encouraged their uncle. Blaine said, "Oh, that's okay, Uncle Cecil. At least that big tree in your backyard with the tire swing is nice and healthy."

"It sure is." Cecil smiled. "And it's not the only plant in this neighborhood that's looking good! Come with me, you two. Let's take our botany introduction on the road. I want you to meet some of the neighbors and see their. . . um. . . much healthier plants!"

Blaine and Tracey followed their exuberant uncle as he skipped down the front porch steps and made his way up the sidewalk. Cecil talked as he walked, using his hands to communicate almost as much as he used his mouth.

"As I said before, botany is the study of plants. You two will learn every blooming thing there is to know about all kinds of different plants, including mosses, ferns, conifers, flowers, and more! The plant kingdom spans all the way from the world's largest living tree, the giant sequoia, to this little tiny weed poking up right here through the crack in the sidewalk." Cecil bent over as he stopped to point at a small green sprout protruding from between two pieces of concrete.

He looked back at each twin and whispered, as if what he was about to say was top secret, "Actually, there are plants even smaller than that."

Cecil stood upright and continued walking down Pecan Street. "Algae and fungi are not part of the plant kingdom, but you will be studying them on this leg as well," he informed the twins as he waved happily to an older woman sitting on her porch at 1106.

She waved back with a smile on her face. "Hello there, Cecil," she called.

"Hello, Mrs. Pascapali! How are you on this fine day?" Cecil asked.

"Doin' right wonderful. Is this the niece and nephew that you talk so much about?" the neighbor responded.

"It sure is, Mrs. Pascapali. This is Train and Blaisey!" The twins rolled their eyes as their uncle messed up their names once again.

"Well, isn't that nice," drawled Mrs. Pascapali as she waved to the twins.

The twelve-year-olds waved back as they followed behind their uncle. Cecil pointed out and named the Japanese maple tree and the geraniums in Mrs. Pascapali's front yard.

As they moved onto 1108 North Pecan Street, Cecil mentioned, "The guy who lives here tends to be a recluse. None of us in the neighborhood see him very much, but you two are not going to believe what he has in his yard!" Cecil clapped his hands in delight.

The twins looked beyond their uncle into the man's yard, but they didn't see anything out of the ordinary. "What?" they asked simultaneously. "What is so special in his yard?"

"He has three *Sassafras albidum*!" Cecil exclaimed with bright eyes.

Blaine and Tracey looked clueless, so Cecil explained, "*Sassafras albidum* is the scientific term for Sassafras tree! See the three different types of leaves on each of the trees? There are three-lobed leaves, two-lobed leaves, and no-lobed elliptical leaves all on the very same branch! This tree doesn't just give us our name. It can also be used to dye fabric yellow. Back in the day, it was also used to flavor tea and root beer, but nowadays, we know that it can cause cancer, so we stick to enjoying the Sassafras tree with our other senses. Take a big whiff. That amazing smell is the essential oils from the Sassafras tree, which are used in perfumes and soaps. Isn't that awesterriffic?!?"

Blaine and Tracey stared in wonder at the trio of trees they saw in the yard. So that was what a Sassafras tree looked like. It was truly beautiful! They hoped they could meet the man who lived here before they left at the end of the summer and talk with him about his gorgeous trees.

Cecil looked both ways before he started to cross the street. "Over on the other side of the street lives Old Man Grusher," the twins heard their uncle say as they followed him across the road. "He has quite a wide variety of interesting plants in his front yard, but there is always something I forget about his house. What was it? I think it was really important. . . but I just can't ever seem to remember. . . maybe it was. . . DOG! Beware of dog!"

Just as Cecil said this part of his sentence, Blaine and Tracey turned to see a curly-haired black miniature poodle bound off the front porch and rush toward them, furiously barking all the way. At the sight of the dog, Uncle Cecil raised his hands above his head, screamed like a choking chicken, and began running toward home. Blaine and Tracey instantly recalled that their eccentric uncle was, for some strange reason, afraid of dogs. He could barely even stand to be around a puppy without starting to shake in fear.

The twins didn't scream though because it was only one little dog. Instead, they turned to follow their blubbering uncle as they kept one eye on the approaching pooch. It was right behind the twins in a flash, but instead of biting them, it ran right past them and went for Cecil. The little barking dog jumped up and chomped onto Cecil's white lab coat as it flapped in the breeze behind him. The dog managed to rip a piece of it off, but it obviously wasn't enough to satisfy Old Man Grusher's mini-poodle. He spit it out, caught up with the scientist once more, and nipped at his bunny slippers. At this new threat, Cecil tried for all he was worth to get home without losing any more of his clothing. By now, both Blaine and Tracey were laughing uncontrollably. It was quite a sight to behold.

Somehow, Cecil managed to get in his front door without the black dog snatching any more of his attire. He rushed forward, slamming the door as he passed, thus preventing the miniature pooch from entering the house. The twins reached the door soon after, and Old Man Grusher's dog gave them a passing glance as it turned and trotted back home.

They opened the door and walked inside to find a trembling but relieved uncle. He plopped down on a living room couch, exhausted from his epic battle with the much-feared pooch. Cecil took a deep breath before managing a crooked smile as he squeaked out, "Now you see why I chose a prairie dog, not a typical dog, to be my lab assistant."

The twelve-year-olds smiled and laughed.

"I'd planned on giving you two a much more in-depth introduction to botany with a longer tour through the neighborhood." The red-haired scientist bounced back. "But instead, how about you Sassafrases just go check out botany for yourselves! Are you ready to zip?"

Blaine and Tracey looked at each other with bright excitement. "Yes!" they responded in enthusiastic unison.

Immediately, the twins got their smartphones out and opened up the LINLOC applications. Cecil smiled as he watched his niece and nephew. He was overjoyed by their enthusiasm regarding their summer science adventures.

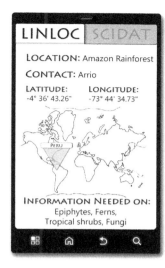

"Well, what does it say?" he asked. "What is the location for your first leg of botany?"

"Peru!" Blaine exclaimed. "We are going back to Peru!"

"Our topics for study are epiphytes, ferns, tropical shrubs, and fungi," Tracey added. "Look, our local expert isn't Alvaro Manihuari this time—it's Arrio!"

"Arrio?" Blaine asked. "Wasn't he Alvaro's Peruvian friend?"

"Yes," Tracey confirmed. "He's the native who helped save us from Ortiz and the illegal loggers. It's strange, though. I don't think he ever spoke one word to us the entire time we were around him."

Blaine smiled. "We're headed back to Peru, starting another scientific subject with a local expert who doesn't talk. This is going to be interesting."

Chapter 2: Return to the Jungle

Falling Orchids

Swirls of rip-roaring light encased Blaine and Tracey Sassafras. Their smiles were almost bigger than their faces could handle. Coasting along the invisible zip lines never got old!

The twins had all but forgotten about the zip-lining summer camp they had wanted to go to just a few weeks before. Granted, Camp Zip Fire was an exciting place, but it couldn't compare in the least bit to traveling all over the globe to learn science on Uncle Cecil's invisible zip lines.

The light-speed travel came to an abrupt halt, and the carabiners automatically unclipped from the zip lines. The twins' bodies slumped into exhausted but exhilarated heaps. As usual, it took a few moments for Blaine's and Tracey's bodies to recover. The twins knew that the blinding white they could see now would slowly fade back into color. All their senses would normalize, and their strength would return after a few moments of tingling and lethargy.

The zip lines were designed to put the twins as close to their local experts as possible without their landings being detected. The twins knew that under no circumstances were they supposed to tell anyone about the existence of the invisible zip lines. So far, during their studies of zoology and anatomy, they had been able to keep the secret.

"Ahhh! Blaine!" Tracey gasped as her sight returned. "It looks like we have landed in the top of a tree!"

"Whoa, you're right, Trace!" Blaine shouted back. "What's the matter? Are you suddenly afraid of heights?"

Tracey shook her head and rolled her eyes. Blaine was always giving her a hard time and acting like an older brother, even though he was older by only five minutes and fourteen seconds.

Blaine stood up and grabbed a branch above him. "I'm just still amazed the zip lines can drop us into such precise spots."

Tracey nodded, agreeing as she stood up as well. "It looks like we are pretty close to the top of this tree. Do you want to see if we can climb all the way up and get a view of the canopy?"

"Sure!" Blaine grinned. "As long as it is a race!"

Tracey jumped into action before Blaine could say Rutherford B. Hayes. She grabbed a branch and hoisted herself up in front of him. He was quick on her heels, and the race was on. Branch after branch, the twins made their way swiftly to the top of the tree. The competition ended in a virtual tie, as the twelve-year-olds raised their heads just above the treetops.

"Wow!" Blaine panted as he pointed out, "Look at the Amazon rainforest! It sure is breathtaking, isn't it?"

Tracey nodded. "It sure is, but you're still the loser."

"Loser?" Blaine objected. "I clearly won that race. You were the loser."

Tracey just smiled and shook her head in disagreement.

Blaine wasn't ready to concede. "What are you talking about? I got my head above the canopy fir—"

A loud snapping sound interrupted the end of Blaine's sentence as the branch he was standing on broke. Tracey saw him disappear down into the green of the tree.

"Blaine!" Tracey shouted. "Where—" Another snapping sound. This time it was Tracey's branch. She felt herself tumbling down through the branches of the tree.

Bump, thud, snap—the Sassafrases went tumbling down, until they both managed to grab a branch. Hanging there by their hands, high above the forest floor, the twins grinned nervously at each other.

Once he caught his breath, Blaine started to speak, "Well, that was a clos—" Before he could finish, his new branch broke, and down he went again.

Tracey watched as her brother managed to grab a vine that sent him speeding away from the tree toward another tree. Blaine then swung down, Tarzan-like, through the rainforest using available vines as they appeared. As he was making his way down, his foot got stuck in a tangle of vines. He reached out to a nearby tree to prevent an inverted dive to the ground, but all he managed to do was grab a nearby flower that was growing on the tree. The flower came loose, and Blaine kept tumbling down with it in his hands. The boy continued to careen down toward the forest floor face-first.

Tracey held her breath as she watched Blaine's crown headed straight for the ground at a fast rate of speed. Just before impact, the vine stretched to its full length, and Blaine came to a stop before bouncing back into the air a bit. Tracey started laughing when she saw that Blaine's tumble ended with him hanging upside-down, caught by his left ankle. His head was at least five feet from the ground, and he was still holding on to the flower.

Tracey swung from her branch to a larger one further down

and then descended the big tree to meet up with her brother. She jumped down to the ground and jogged over to where Blaine was hanging upside down. She reached him in no time and was about to zing him with her sarcastic wit when she saw something that made her stop in her tracks.

A native man wearing only a satchel, a dagger, and a loin cloth was standing completely still, staring directly at Blaine. As Tracey approached, the man's body did not move, but his eyes shifted from Blaine to Tracey and then back to Blaine.

Blaine's breathing stopped, like he was holding his breath, and then he suddenly burst out in exclamation. "Arrio! Trace, look! It's Arrio!"

Tracey looked more closely at the man who had some kind of red paint smeared in lines across his face. Blaine was right! She recognized the man. It was Arrio.

Still hanging upside-down, Blaine attempted to explain to their new, yet familiar, local expert what had just happened. He used big hand motions and loud, slow English. "We . . . fall from sky . . . land in tree . . . me . . . won race . . . top of tree."

Tracey again rolled her eyes at this part of her brother's explanation. Blaine continued. "Then . . . I fall from tree . . . tumble through branches and vines… me also win race to bottom of tree . . ." Blaine looked at Tracey and smiled. Tracey just sighed, not impressed.

Arrio's expression changed from a non-expression to one with a slight smile. "Why are you talking so funny, Blaine?" the native Peruvian said in perfect English.

Blaine's face now held a look of confusion, even viewed from an upside-down angle. "What? We thought . . ." the boy stammered.

"We didn't know you spoke English," Tracey answered for her brother.

"You never asked," Arrio stated plainly. "What are the two

of you doing back in the Amazon?"

"We want to know more about the plants of the rainforest," Tracey answered. "Can you help us with that?"

"I would be happy to," replied the tribesman. "I don't consider myself an expert like Alvaro, but I have been working to learn as much as I can. My ancestors have passed down thousands of years of knowledge about the rainforest to me and my fellow tribesmen. You know that rainforests are found all over the world, but the largest is here in the Amazon."

Arrio unsheathed the dagger from his side and swiftly cut Blaine down. He landed with a thud and jumped up as quickly as possible, flower still in hand. Arrio reached out, grabbed the plant, and looked it over.

"What you have, Blaine, is a Moth Orchid. It is one of more than twenty thousand species of orchid found here in the Amazon rainforest. Orchids are epiphytes, which means that they are plants that sprout and grow on the branches of trees. Generally, these plants do not harm the tree to which they are attached. Instead, they use their position in the tree to receive more light than the other plants on the forest floor. The orchids manage to get the water and nutrients they need from the air and the rain that comes down through the canopy. Epiphytes include orchids, ferns, and bromeliads, all of which can be found in tropical climates. Mosses and lichens, which are normally found in more temperate climates, are also considered epiphytes."

The Sassafras twins were amazed. Not only did Arrio speak, but he also spoke English. And not only did he speak English, but he also spoke it well enough to clearly explain what an epiphyte was. For someone who didn't consider himself much of an expert, Arrio sure knew a lot of detailed information about the orchid.

Both twins smiled as they used their phones to take a picture of the Moth Orchid. They were glad Peru was on the itinerary again, and they were ecstatic that Arrio was the local expert. They were looking forward to getting to know him more.

Arrio continued. "Orchids are known worldwide for their fragrance and beauty. However, there is so much more to these amazing plants. Did you know that their seeds are carried throughout the rainforest by the wind? Or that they can survive with little water? In the rainforest, orchids typically attach themselves to trees and exposed roots of trees, so they can collect as much water and nutrients as possible."

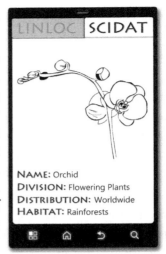

NAME: Orchid
DIVISION: Flowering Plants
DISTRIBUTION: Worldwide
HABITAT: Rainforests

The native handed the orchid back to Blaine as he went on. "Personally, my favorite orchid is the Cattleya orchid. It has a thickened bulb at the base of the stem and wide fleshy leaves that store food and water. Cattleya are usually found on trees near a stream or higher up because they prefer a fair amount of sunlight. The stream allows for a break in the canopy, which lets more sunlight in."

Arrio paused and looked curiously at the twins before asking, "What were the two of you doing up in that Kapok tree?"

"We were trying to get a peek above the canopy," Tracey answered.

"And we were racing," Blaine added. "A race that, of course, I won."

Tracey elbowed her brother in the ribs. "I won the race," she stated confidently. "And besides, I'm not the one who fell all the way down."

A slight smile again formed on Arrio's face as he witnessed the Sassafrases' good-natured competitiveness. "Well, you just encountered the five layers of the rainforest. Of course, Blaine did so from the top down. He fell first through the emergent leaves, then the canopy, next the understory, then the shrub layer, and finally, he nearly hit the forest floor. This Kapok tree you two were in is the most important tree species in the rainforest. It plays host to many of the Amazon's plants and animals. It is also where my tribe, the Yora tribe, and others find the bark, resin, seeds, and leaves that we use to make medicines we need."

Arrio paused and opened up his satchel. "But not even the Kapok tree and all of its wonderful resources can cure every disease. That's why Alvaro just gave me this," he said, pulling out a vial of medicine.

"What is that?" Blaine asked.

"This is the measles vaccination," the native said. "I was just now heading back to my tribe to administer it to those who need it. Do you two want to come with me?"

The Sassafras twins nodded vigorously. The twelve-year-olds followed the native Peruvian on a narrow and heavily foliaged trail through the rainforest. They were happy to follow Arrio and help with anything he or his tribe needed. One thing did worry them, though.

As they thought back to their first trip to the Amazon, they remembered how Arrio's tribesmen had helped rescue them from some nasty illegal loggers. The tribesmen had used blowguns in the process, blowguns that had been equipped with poison darts coated with a toxic goo secreted from the Amazon's poison dart frogs. The twins guessed that this meant Arrio's tribe was a warrior tribe that was probably not very used to outsiders. Sure, he was nice and they could communicate fine with him, but what would happen when they marched into the tribal camp unannounced? The twins gulped and prayed that Arrio's tribesmen were as friendly as he was.

They had been hiking for over two hours when finally the

three came to a small clearing. Arrio walked purposefully out from the cover of trees with the twins cautiously following behind him. At first, the only thing Blaine and Tracey saw in the clearing was a cluster of small huts. Then, slowly, dozens of the Yora tribe members made their way out into the open to greet Arrio and his curious guests.

The men were dressed just like Arrio, wearing only loincloths. They had red paint streaked in different places across their faces and bodies. They had piercings in their ears and noses, mostly ornamented with bone. The women also had multiple piercings, but they were even more heavily adorned. They wore multiple necklaces and bracelets, as well as earrings and nose rings. They had the same red paint, but it was mainly concentrated around their eyes and cheekbones. The women's clothes covered more of their bodies than the men's did, but everyone was barefoot. In addition, they all seemed to be holding some kind of weapon.

The Sassafras twins stood nervously still as the Yora surrounded them. They examined the twins by grabbing the twins' arms, squeezing and pushing at different parts of their faces, and running their hands through their hair. The twins glanced over at Arrio to see his reaction. The man wasn't too expressive, but the twins were pretty sure that he was laughing. The twins relaxed a bit because that meant this behavior from his tribe mates was customary.

After their curiosity was satisfied, the natives went back to doing the various tasks they had been doing. The twins, who were now apparently accepted by the tribe members, joined Arrio as he went around administering the measles vaccination.

"Here in the Yora tribe," Arrio told them, "we have been visited by several foreigners before. So far, all of these foreigners have been kind and helpful. For example, Alvaro Manihuari, whom the two of you know, has brought medicine several times for us. He also made the effort to learn our language, on top of teaching me Spanish, English, and Latin. Most important of all, he has taken up the cause to protect our native land from the illegal intrusion of loggers and oilmen. Over the years, he has become a very good friend of mine. We are always willing to give each other assistance and teach each other about our different cultures.

"However, this has not always been the case for the other native Amazonian tribes of Peru. Many know foreigners only as strange-looking people who steal their land. Foreigners have destroyed their hunting and farming grounds, and spread deadly diseases. So many native tribes have become suspicious of and violent against these invaders." Blaine and Tracey both shuddered. They hoped they never came across any of these violent tribes.

"It is true the shrinking rainforest has become a major problem." Arrio continued. "Much of this is natural and can't be stopped, but when oil companies and logging companies come to our forest and illegally cut, dig, drill, and destroy, they escalate the problem and threaten our very livelihood. Tonight, around the fire, we are going to have a tribal meeting about this very subject."

The Sassafrases tried to be as helpful as possible as they assisted Arrio in getting everyone vaccinated. Their minds dwelled on the things he had said about the shrinking rainforest and the intrusion of foreigners. They wondered if there were any solutions to these problems.

Fishing for Ferns

When the vaccinations were finished, Arrio invited the twins down to the river to try their hands at fishing with a bow and arrow. Always up for a challenge, both Sassafrases went with him. Arrio took several steps out into the shallows of the river and placed an arrow on his bow. In a matter of seconds, after just one shot, he got a fish. Now it was the twins' turn. It took a while, but what seemed like an impossible task at first was actually just a matter of standing in the water completely still and keeping steady aim. After a bit of effort, Blaine and Tracey managed to catch a fish.

Blaine, who took his turn after Tracey, handed the bow back to Arrio. He caught several more fish for dinner much faster than the twelve-year-olds. As Arrio fished, he pointed out some of the plant life along the riverbanks to the twins.

"Do you see those big green plants with the tapered leaves?" Arrio asked his honorary tribe mates.

They nodded, and he responded, "Those are ferns."

The twins immediately snapped pictures before Arrio continued. "Ferns are non-flowering plants that grow in damp or shady places like humid forests or river banks. They have delicate tapered leaves called fronds, which unfurl as they grow. Ferns have rigid stalks that transport nutrients, and their roots are typically small. So, they do not go deep into the soil."

NAME: Fern
DIVISION: Simple Plants
DISTRIBUTION: Worldwide
HABITAT: Woodlands and Rainforests

Swip! Arrio let an arrow go and caught another fish.

"Here in the Amazon," he went on, "ferns can grow on the forest floor or in trees. The filmy fern is an epiphytic fern that can take root in the boughs or trunks of a tree. These ferns absorb nutrients from dead insects, leaves, and droppings that accumulate around their roots."

Swip! Another fish.

"Flowering plants, like the orchids that we talked about earlier, use their flowers to reproduce. But non-flowering plants, like these ferns, use spores to reproduce. In other words, instead of blooming flowers, these plants release spores. These microscopic particles of living material are encased in a tough coating. They develop in sacs called sori on the underside of the fern fronds. When the spores mature, they are released by the thousands into the air. When the spores land, they reproduce a copy of the fern plant from which they came."

Swip! Swip! He shot two more fish in a row.

The twins helped Arrio carry all the fish back up to the clearing, where they sat down with some of the women of the tribe, who showed them the proper way to gut and clean their catch. When the fish were ready, they were thrown onto a fire. Once they were done, they were served as dinner to the entire tribe, along with some other unfamiliar vegetables that had already been prepared.

Blaine and Tracey sat on stumps around the fire and enjoyed their dinner and the company. "How many people get the chance to spend time with indigenous Amazonian tribes?" the twins thought to themselves as they watched the interaction and camaraderie of the Yora tribe.

After dinner, as the forest grew dark because of the sinking sun, the elders remained seated around the fire and began to discuss the grave issues their tribe was facing. They talked about what to do about foreign threats and their shrinking homeland. The Sassafrases listened to the Yora language being spoken as different men from the tribe took their turn sharing their opinion about what should happen. Some were very animated and loud, and others spoke more slowly and calmly. Many of them pointed toward Arrio as they spoke, giving the indication that he played an important role in the tribe's decision.

In the end, all the men seemed to agree. They got up one by one and left the fire area to retire for the night, leaving only Arrio and the Sassafras twins sitting by the crackling orange embers.

Blaine and Tracey couldn't tell by looking at the native man's face if he was happy with the decision the tribe had made. The twins were expecting Arrio to share with them what that decision might be, but instead he asked, "Do the two of you want to hear some traditional Peruvian folklore?"

"Sure," Blaine answered.

"I love campfire stories!" Tracey added, smiling.

Arrio added a couple of logs to the fire and then began. "It is said in Incan tradition that the sun god had eight children—four sons and four daughters. He sent his children to Earth and put them in a cave. One day, the eight of them made their way out of the cave. Manco, the oldest son, held in his hand a golden staff. He told his brothers and sisters they would walk across the Earth, and wherever this golden staff sunk into the ground that is where they would start their civilization. The four sisters, named Ocllo, Rana, Huaca, and Cora, all agreed that this was a good plan.

"Manco's brothers, however, disagreed with his plan. They each had plans of their own. Cachi, the second oldest son, said it was by strength and power that they should start a civilization. Cachi bragged that he was the strongest and most powerful of all the brothers. Manco, being very tricky, asked Cachi, 'Since you are the strongest and the most powerful, will you go catch a sacred llama that I saw back in the cave?'

"Cachi, feeling flattered, went back into the cave to look for the sacred llama. As soon as he went in, Manco and his sisters rolled a huge stone over the cave's entrance, blocking Cachi inside forever. Uchu, the third oldest son, quickly said he did not want to walk with the golden staff until it sunk into the ground. Instead, he wanted to climb on top of the cave and look over his brother and sisters and the civilization they would start. As soon as Uchu climbed to the top of the cave and sat down, he immediately became a pillar of stone.

"Auca, the fourth oldest son, then said he didn't want to walk with the golden staff. Instead, he wanted to wander off on his own and start a civilization by himself. This is what Auca did, and neither Manco nor his sisters ever saw Auca again. Manco and his four sisters then walked for a very long time until the golden staff finally sank into the ground at a place called Qusqu. That is where Incan civilization began, and today, that place is the modern

Peruvian city known as Cuzco."

Arrio paused and looked at the twins. "The rest of the story turns from folklore to historical fact, but it is quite sad. Do you want to hear this part of the story, too?"

The twins cautiously nodded.

"The Incans prospered and grew until they became a great civilization, but even great civilizations are not beyond the grasp of tragedy." Arrio's face was grave. "In the fifteen hundreds, Spanish conquistadors came to the Amazon by the thousands. Their arrival led to the demise of the Incan Empire. The powerful conquistadors' lust for land and gold led to the destruction of the Incas, but most devastating of all were the disease and sickness that the Europeans brought across the Atlantic Ocean with them. Measles, influenza, and smallpox wiped out more than ninety percent of the Incan population because their immune systems could not fight these newly introduced diseases. Many of the remaining Incans fled deep into the heart of the Amazon rainforest. Today, most, if not all, of the untouched native tribes of the Peruvian Amazon have Incan blood running through their veins." Arrio stopped, took a long deep breath, and then frowned as if thinking about the struggles his tribe was facing.

"It is true the world needs oil and wood," he told them. "There is nothing wrong with using the Earth's natural resources, digging for oil, or harvesting logs, when it is done wisely and responsibly. These oilmen and loggers who are illegally encroaching on our beloved forest with no regard for those of us who live here or for anyone who will come after them are a tragedy. If nothing is done, they will be like another wave of conquistadors."

As they listened, Blaine's face remained somber. Tracey's eyes welled up with tears as Arrio continued. "These oilmen and loggers, now in the twenty-first century, have brought with them some of the same sicknesses that the conquistadors brought back in the fifteen hundreds, and our immune systems are still too unfamiliar with these diseases to fight them. With the help of people like my friend, Alvaro Manihuari, we can protect ourselves against this problem and the other problems that threaten our existence."

The Sassafrases stared into the fire and remained silent as they thought about all their friend had just said. They definitely wanted to be part of the solution and not part of the problem. Arrio stoked the coals of the fire and then looked back at the twins.

"Would you like to hear another Yora legend?" he asked, the flames casting an orange glow on his face. The Sassafrases nodded, wanting to turn their thoughts to more pleasant avenues.

"The legend I'm now about to tell you is not widely known in Peru or any other place. It is circulated by oral tradition only, exclusively among the indigenous tribes of the Peruvian Amazon." With that mysterious introduction, Arrio already had the twins' attention. "As I said, the four brothers and the four sisters that I mentioned earlier were said to all have been children of the sun god. When Manco and his sisters founded Cuzco, they built a temple there for their father named 'The Temple of the Sun.' Legend has it that while Manco stayed in Cuzco and ruled over the city and the temple there, his four sisters spread out throughout the jungle, and each built smaller sun temples of their own.

"These smaller temples were said to be about the size of small huts and virtually impossible to open because they were built like puzzles. In these miniature sun temples, the sisters each put pieces of Manco's golden staff, which they had taken with them. Legend says those small pieces of the golden staff could be used by someone with Incan blood again to point the way to a location that could become the start of another Incan civilization." Arrio paused and looked into the twins' wide, wonder-filled eyes.

"The location of these smaller sun temples has remained a mystery for hundreds of years. Not even during the invasion of the Spanish conquistadors were they found, but all of that changed two days ago."

"What?" Blaine sputtered. "What changed?"

"Two days ago," Arrio answered slowly, "a group of loggers unearthed one of the sun temples—The Huaca Sun Temple."

Now it was Tracey's turn to gasp, "What? How is that possible? You're telling me that after hundreds of years, illegal loggers were the ones to find one of the miniature sun temples?"

"No, no," Arrio answered. "It wasn't illegal loggers. These were ProLog loggers, who are perfectly ethical and responsible. They work for Ernesto Perez."

"ProLog!" Blaine exclaimed. "They are the ones who cut us down out of a tree, captured us in a net, and threatened to cut down every tree in the Amazon!"

"Not exactly," Arrio corrected. "It was the foreman Ortiz and several rogue ProLog loggers who did all that. They have since been locked up in Iquitos and are currently awaiting trial. Ernesto Perez made it clear that Ortiz and the rogue loggers were not representing him or ProLog when they did those things."

The Sassafrases thought back to the time they had spent with Ernesto Perez and his children, Vancho and Violetta. The three of them had all been very friendly, and Ernesto had seemed like a good and honest man to the twins.

"So, what is going to happen to the temple?" Tracey asked.

"Ernesto Perez has already put down the money to have the temple excavated, restored, and then carefully transported to Cuzco, where it will be put on display in an Incan museum." Arrio's eyes were bright with excitement.

"But what about the piece of the golden staff that was inside the temple? Do you think it's still there?" Blaine was full of curiosity.

Arrio responded, "Before I tell you more about that, may I change the subject?"

"Of course," the twins chimed together.

The Yora man stoked the fire once more and then started again. "Earlier, the elders of my tribe were discussing most everything we've just talked about. Different elders have different opinions on what our current course of action should be. Some want to fight, some want to flee deeper into the forest, and a few even thought we should try moving to the city. In the end, they decided we should stay where we are and work through our problems with the cooperation of wise outsiders, like Alvaro. During the discussion, though, another issue came up."

Arrio turned from the fire and looked out into the dark forest. "There is another tribe out there. They call themselves the Matsigenka tribe."

"The Matsigenka tribe?" the twins repeated, looking deep into the forest.

"Yes," Arrio confirmed. "One of the elders heard through the Kapok vine that the Matsigenka may be planning an attack on Ernesto Perez and the archaeologists he brought."

"An attack?" Tracey gasped.

Arrio nodded. "The Matsigenka, who are led by a fierce warrior named Itotia, have reacted differently to the threats than our tribe. They have a history of violence against foreigners. When they heard one of the legendary miniature Incan sun temples had been found by a group of outsiders, they became angry. They believe in the magic of the golden staff, though it is only a legend. Itotia is probably now planning to attack the outsiders, and then find a way to open the Huaca Sun Temple. He hopes to take the section of the golden staff that may or may not be inside, and then use it to guide them to new land."

The native looked back at the fire. "But back to your question, Blaine. The one about whether I think there really was a golden staff and if a piece of that staff is inside the Huaca temple." Arrio paused. "I just don't know. What I do know is that Ernesto and his men may be in trouble. And I know that the Yora have never gotten along with the Matsigenka. Our tribes have had conflicts with each other for what seems like forever. For generations, they have used their spears against us, and we have used our blowguns against them. Over time, we have become a tribe that prefers peace and cooperation, but not the Matsigenka. They always lean toward violence and strife, and so they remain one of the most feared tribes of the Amazon. One of the things they're most infamous for is the practice of 'wife stealing.'"

"Wife stealing?" Tracey asked, alarmed.

"They sneak into another tribe's camp and kidnap as many of the women as they can," Arrio explained.

Arrio stood up from his fireside seat. "Blaine and Tracey, I can tell by the looks on your faces that I have given you more than enough to think about. It will do us all good to now get some rest. Let me show you where you can sleep."

The twins got up and followed Arrio to one of the communal huts, where he led them to a couple of empty hammocks. Not five minutes later, both twins were fast asleep, swaying gently in their hanging beds. One thing about traveling by light-speed zip lines is that they definitely made these twelve-year-olds tired.

Since the Amazon rainforest was so dark, he was pretty sure he didn't even need this invisible Dark Cape suit to sneak up on the twins. He was using it now anyway, just to be safe. He still hadn't become completely accustomed to being invisible. It was a little strange to feel your body doing something but not be able to see it.

His new sabotage plans were centered on the twins' smartphones. Really, those phones were the key to everything. It was how they recorded the data they were learning, realized their scientific locations, and communicated with their uncle. He took a quiet step toward their sleeping quarters. He looked around the hut at the various hammocks.

There they were—in two hammocks right next to each other. They looked like they were sound asleep, as did all of their local Amazonian friends. To make matters even better, the twins' backpacks were on the ground below their hammocks. He was sure that's where they kept their phones. A wicked grin formed on his invisible face. This was going to be easier than taking candy from a baby.

Chapter 3: Kidnapped

Lilies on the Trail

Tracey didn't know if she was awake or dreaming. She felt as if something, or maybe someone, was silently approaching her. She tried to push the jumbled pieces of floating thoughts in her brain together to form comprehensible recollections.

Where was she again? Was it Beijing, or maybe Dubai? No, she was in Peru—sleeping in the Amazon rainforest in a hammock. Who was approaching her? Was it Blaine? Was it Arrio?

Tracey slowly blinked open her eyes. It was dark. She waited a moment for her pupils to adjust. She could see that she was in a hammock in a hut, but who or what was approaching her? She couldn't see anything moving. She strained her ears for any sounds. Nothing.

She saw nothing. She heard no sound, but she still had the feeling that something was there. It felt like an invisible form was sneaking up on her.

Suddenly, Tracey felt her body being lifted out of her hammock. She tried to scream, but a big hand immediately clamped down over her mouth. She kicked her feet and reached out, grasping for something, anything at all. It was of no use. Whoever or whatever had her was as strong as a bear and as stealthy as a fox. Tracey's wide eyes watched as the hut where the Yora tribe and her brother Blaine still slept disappeared as she was carried off into the dark jungle.

"Yowzers!" That is not a word that came to his mind very often, but that was the only word he could think of right now. He had been crouching down, just about ready to unzip that girl's backpack. He was going to steal her phone. Suddenly, out of nowhere, the biggest tribal man one could imagine had flashed into the hut. In one swift, silent move, he had swiped her from her hammock and had carried her off into the night.

"Yowzers!" he thought again. "That was scary!"

He could feel his invisible body shaking. What if that tribal man came back? What if the man had bumped into his invisible form and became angry? What if he had used that spear that he was carrying? The thought made him shudder. He wanted to steal the smartphones and stop those twins and their uncle, but it looked like that may have to wait until the next location. He didn't want to get hurt or kidnapped by that tribal man. That was the biggest, scariest guy he had ever seen. He tiptoed out of the hut and prepared to zip back home. He would just wait to get those twins at the next location. If they even made it to the next location.

"Blaine! Blaine!" The twelve-year-old Sassafras boy heard his name being called and his body being shaken.

"Tracey, why does your voice sound like a man's?" was Blaine's first thought as he opened his eyes. Only, it wasn't Tracey that he saw. It was Arrio, and he looked worried.

"Blaine!" the Yora man screamed again. "Your sister! She's gone! She has been taken!" Blaine sat up straight in his hammock. It was still dark outside, though signs of morning could be seen.

"Taken?" Blaine asked sleepily, "What do you mean?"

"I mean she has been kidnapped!" Arrio replied with an

unusual amount of angst in his voice. "It was the Matsigenka. Quick, get out of your hammock and put your shoes on. We are not sure how long ago she was taken, so if we want to hang onto any chance of getting her back, we have to leave right now!"

Blaine jumped out of his hammock, put on his shoes, and snatched up both his and his sister's backpacks in record time. He ran after Arrio and joined the members of the Yora tribe who had already gathered outside the hut.

"Tenyoa here is an expert tracker," Arrio stated as he pointed to one of his tribe mates. "He will lead the way, but we must do our very best to keep up with him."

Blaine nodded in understanding, full of determination to do just that. Arrio signaled to Tenyoa, who turned and bolted off like a jaguar into the forest.

Blaine had always considered himself a pretty good runner. He had medaled in most of the track events at school, but he wasn't even close to as fast as the Yora men were. They ran through the dark overgrown jungle like it was a flat track in the daylight. He was doing his best to keep up, but this was going to be different. He just hoped he could keep the pace to stay within eyeshot of his native friends. Thankfully, the day was getting brighter with every passing minute as they followed a small trail, over the rainforest floor that was woven with tangled roots and plants.

As Blaine ran on, his mind rolled through a series of questions: Who had taken Tracey? Was it really the Matsigenka, or was the Man with No Eyebrows back? Was she safe? Had she managed to escape? He hoped they could find her soon!

Tenyoa continued to swiftly track whoever had taken Tracey. The line of racing Yora men kept up with his pace. Arrio was second to last, and Blaine brought up the rear. They zigged to the left and zagged to the right. They jumped over a root here and ducked under a branch there. Finally, they came to a place where the trail grew a little wider. Instead of running straight on through, Tenyoa and the Yora men jumped off to the right and barreled over a nonexistent path that ran through heavy shrubbery. Blaine was puzzled by this. He lost track and took a single step beyond the point where the Yora men had jogged off to the right. Immediately, he wished that he had been paying better attention. He felt something tighten around his left ankle, and before he knew it, he was being pulled upside down, up into the air.

"Oh Blaine, c'mon!" The Sassafras boy internally scolded himself. "Why didn't you follow the Yora? Now look what you've gotten yourself into!" He still could not see exactly what he had gotten himself into, because he was swinging around rather wildly. Eventually, he slowed, and what he saw made his stomach sink, or maybe it rose since he was upside down. He was hanging by his ankle over a big pit full of spikes.

She was struck by the man's strength. She was slung over

his shoulder like a sack of potatoes, held by one arm. In the other arm, he held a spear. Tracey wondered, "How is he still able to run through the forest like this?"

He was as fleet-footed as a gazelle. He ran over, around, and through the obstacles of the rainforest as if they weren't even there. One moment she had been sleeping in her hammock, the next she had been in the grasp of this man. He had taken her out of the hut into the darkness. He tied her up, and then he had begun running and had not stopped. She despised him for kidnapping her, but she was amazed at his strength.

After what seemed like hours, the man finally slowed down. His run changed into a jog and then eventually into a walk. He stopped abruptly and took Tracey off of his shoulder, dropping her on the ground. Tracey wondered why, until she flopped over on her other side and saw that a whole tribe of people were there to greet him and her, his prize. Their faces were streaked with black paint in fierce patterns. They each held a spear in their hands. They looked like him, but he was the biggest by far. He stepped toward them as they reached out and congratulated him, oohing and aahing over her.

Tracey lay tied up on the ground recalling something that Arrio had said the night before. Her stomach sank as she remembered the last story he told. Was this the Matsigenka tribe? Was she their newest stolen wife?

"It's a what?" Blaine asked.

"A snare," Arrio repeated. "You have been caught in a snare."

"Well, how do I get down?" the Sassafras boy asked. Arrio looked at Blaine hanging there upside down by his ankle, and then looked at the sizable spike-filled pit underneath him.

"I don't know," Arrio finally replied. "The snare is more complicated than I have seen before."

Blaine groaned. That wasn't what he wanted to hear.

"My guess is that Itotia and the Matsigenka designed this to catch us if we followed him." Arrio continued.

"Well, it sure worked like a charm, didn't it?" Blaine quipped, annoyed at the bad guys' ingenuity.

"It sure did," Arrio confirmed. "Now I can't reach you without falling into the pit myself. Even if I could reach you, I couldn't just cut you down, because if I did . . ."

Blaine looked at the multiple spikes in the pit below him. "Because if you did, it wouldn't be a very good landing."

Arrio nodded gravely.

Tenyoa and the other Yora had not seen Blaine get caught in the snare, so they hadn't stopped. Presumably, they were still hot on the trail of Tracey and her captor. Arrio was the only one with Blaine, so it was up to him to figure out how to rescue the boy.

Blaine looked down and could tell that Arrio had a plan because the native man scampered off quickly into the forest and came back with an armful of small vines. Arrio started braiding the vines together, but instead of talking about the plan, he began telling Blaine about a patch of shrubs that was growing next to the pit.

"This is a peace lily," he said. "You can tell because it has a white flower with a yellow or green stalk in the center called a spathe. It has large green leaves that do not need a large amount of light or water to survive. The peace lily is a good example of the tropical shrubs that are found in the shrub layer of the rainforest."

Arrio looked at Blaine's red face as he asked, "Do you remember yesterday when I told you about the five layers of the rainforest?"

Blaine nodded that he did.

Arrio, still quickly weaving vines together, added, "The air in the shrub layer is moist, hot, and still. It is the layer that has the densest growth. It contains shrubs, ferns, and other plants that need less light. You will notice that the plants in the shrub layer have large leaves so they can absorb as much of the sunlight as they can get."

Arrio looked back toward the flowering shrub before saying, "The peace lily is actually not a true member of the Lily family. Instead, it is a member of the Araceae family. It contains calcium oxalate crystals, which can cause skin irritation, a burning sensation in the mouth, difficulty swallowing, and nausea. Even so, it is used today as a common houseplant."

Blaine was glad that he now had the information he needed regarding this tropical shrub. However, it was a little difficult to get too excited about scientific data when his whole body was going numb. Plus, he was still hanging over a pit full of spikes. Luckily, Arrio was just finishing up his braiding job.

Arrio looked up, finally ready to reveal the plan. "OK, Blaine, I need you to catch this makeshift jungle rope and hold on tight. I'm going to pull it over and tie it around that tree. If you keep hold of it, the rope should pull you out from over the pit. While you're holding on, I will race back and cut you free from the snare. Then you should fall to safe ground. Got it?"

"Got it," Blaine answered, sounding more confident than he really was.

Arrio tossed Blaine the rope, and he caught it on the first try. Arrio then sprang into action and pulled Blaine out from over the snare. He then cut the boy down, and Blaine landed on the ground beside the spike-filled pit instead of inside it.

Arrio put his knife back into its sheath before helping Blaine stand. "You okay?" he asked the twelve-year-old.

Blaine nodded that he was, even though he still felt a little wobbly and his left ankle was sore.

"All right, then let's see if we can catch up with Tenyoa and find your sister."

Blaine nodded again as he used his phone to take a picture of one of the peace lilies next to the pit. He hoped he could share the image with Tracey, as well as the other data he'd collected. In order to do that, they had to find her first.

The night was gone. The light of day had fully arrived, but Tracey Sassafras's predicament had not changed. Once more, she found herself, still tied up thrown over the shoulder of this hulk of a man. She was being carried at a wild pace through the jungle. They stopped briefly when her captor had met up with more of his tribe mates. They had given him their approval of her. Then one of the tribe members had said something very loudly and emphatically in their native language as he pointed deeper into the forest. The big man immediately picked Tracey back up and ran in that direction.

As Tracey bounced around on the man's cannonball-sized shoulder, she again wondered if this was the Matsigenka tribe Arrio had talked about last night. If it was, then Tracey would be willing to bet this big man carrying her was their warrior leader, Itotia. A wave of emotion flushed over Tracey, and she choked back tears. She didn't want to become a part of the Matsigenka tribe. She wanted to be reunited with Blaine and continue their science-learning journey.

She hoped Arrio and the Yora could figure out a way to find her and get her back.

Eventually, Itotia and the Matsigenka—if that is indeed who they were—slowed from a run to a stealthy creep. They began to behave as if they were sneaking up on something or someone. Itotia dumped Tracey on the ground as he and his tribe mates stopped and crouched behind a cluster of ferns. With wide eyes, they spied on something in a clearing past the foliage. Tracey wiggled around a bit and managed to lift her head up enough to get a peek through the ferns.

Right away, she saw it, and she immediately knew what it was. There, on the other side of the clearing, stood the Huaca Sun Temple! As she had been told, it was the size of a small hut. The building was an exact miniature replica of the original Incan temple that had been built in Cuzco, Peru. Arrio's stories were true!

Tracey saw several people tarrying around the structure. Some seemed to be examining it. She guessed they were the archaeologists she had heard about. Other men wearing blue ProLog jumpsuits, were carrying and working with shovels. There were also two middle-aged Latino faces that Tracey recognized—Alvaro Manihuari and Ernesto Perez! They stood a few meters from the temple and looked at a map or blueprints together.

When she saw the two men, Tracey's first inclination was to scream for help, but she stopped herself. What would Itotia and the Matsigenka do to her if she screamed? What would they do to Alvaro and Ernesto? Tracey looked at the tribe members who were spying on the outsiders, whispering to each other, and nervously clutching their spears. She had a very bad feeling about this.

"I wonder if they are about to—" But before the Sassafras girl could even finish her thought, Itotia, the big man who'd kidnapped and carried her, let out a long fierce holler. He lifted his spear above his head and ran into the clearing. His fellow tribesmen joined him with shouts and spears, running behind him as a bold group toward the unsuspecting foreigners.

Fighting Fungi

Alvaro and Ernesto, along with their men, looked up with horror in their eyes. Some stood still, shocked and at a loss for what to do. Others, including Ernesto Perez, quickly reached out and grabbed nearby shotguns. The fierce natives continued their rush forward until they saw the guns. They stopped in a line facing the outsiders, still shouting and still holding their spears high. The loggers and archaeologists stood their ground, aiming their guns and shouting warnings.

Both sides stood. Both sides shouted. Both sides threatened. It was a quivering, precarious standoff that could be snapped into something more violent with one wrong move on either side.

Though both her hands and feet were bound, Tracey managed to stand up by using the trunk of a tree close to her. "This is not good," Tracey thought. "Not good at all."

Tracey, of course, didn't want the Matsigenka to spear the foreigners, but she also didn't want the foreigners to shoot the Matsigenka. Nothing good could come out of this; she had to do something.

Suddenly, in the middle of the fray, Tracey heard a strong clear voice. It was Alvaro Manihuari. He had walked out into the middle of the clearing and was standing alone between the two opposing lines.

"Everyone, lower your weapons," he commanded with hands raised toward both sides. He then looked toward Ernesto and his men.

"Ernesto!" Alvaro pleaded to his friend. "These are the very people we were just talking about! They are why your company conducts responsible logging operations. They are who we are preserving this archaeological find for. They are one of the indigenous tribes with Incan blood for whom we are trying so hard to save the Amazon rainforest. Put your weapons down!"

Not a muscle moved in Ernesto's body except the ones in his mouth. "I agree, Alvaro, but I didn't know they would be charging us with spears!"

Ernesto's gun remained aimed, as did all the guns of his men. "You know I don't want to shoot anybody." The ProLog president continued. "Especially these remarkable native people, but I must protect myself and my men."

Alvaro nodded, understanding, but kept a restraining hand up in the logger's direction. "OK, Ernesto, but please just don't do anything rash. Let me see if I can use some of the language that I have learned to communicate with them."

Manihuari then slowly turned toward Itotia and the Matsigenka and cautiously tried out some native words to see if they understood. As Alvaro spoke, the tribal men looked at each other, obviously surprised that this foreigner could speak some of their language. However, like the men facing them, they did not move their aimed weapons. After Manihuari labored through a few minutes of the native tribal tongue, Itotia interrupted him and began speaking for his tribe. Tracey could not understand a word of it, but there was no questioning that it was direct and authoritative.

When Itotia was finished, Alvaro turned toward the loggers to attempt a translation, but before he got two words out, the Matsigenka began moving forward with spears in hand. The loggers all stood up straighter and took nervous steps back. Each man peeked at Ernesto to see what he would do. Alvaro Manihuari remained standing in the middle of it all, with his hopes of making peace quickly diminishing. Tracey's heart pounded. She had to do something, but what could she do? She was neither a warrior nor a logger, and she was all tied up.

With no better plan coming to mind, Tracey bounded out into the clearing clumsily and screamed as loudly as she could,

"Stop! Put down your guns and spears!"

She was about to shout it a second time, but she only got out "sto—" because she tripped and fell to the ground face-first. The landing hurt, but what stung more was how badly her attempt at helping had just gone.

Tracey's moaning would have continued if she hadn't heard footsteps coming her way. She felt a big hand pulling her up off the ground. It was Itotia, and he wasted no time in throwing her back over his shoulder. Tracey wondered if she was going to be bounced back through the forest or carried into battle or some other unpleasant fate.

Itotia shouted as he lifted his spear high. With Tracey over his shoulder, he began leading his tribe forward. Again, a clear strong voice interrupted the would-be battle.

"Tracey Sassafras?" Alvaro exclaimed. "Is that you?"

Tracey managed a smile that said, "Yes, I'm sorry, I'm embarrassed this is happening, and can you please rescue me," all in one.

Manihuari started talking to Itotia and the Matsigenka again in their native tongue. Tracey couldn't understand what Alvaro was saying, but she could tell he was pleading with the natives to stop and consider peaceful options.

The Matsigenka continued to inch forward with spears raised.

Ernesto and his men stood their ground with weapons aimed.

Alvaro Manihuari continued to plead, contending for a peaceful solution. The volume of his pleading got louder until he was literally shouting.

Then, something he said struck a chord with the native tribe. They stopped and all stared at Manihuari as if they were wondering if what he had just said was true. Seeing that he had their attention, Alvaro said a few more ardent words in the local language. Amazingly, Itotia and the Matsigenka all lowered their spears.

Alvaro quickly turned toward the loggers, smiling. "Ernesto, my friend, we have reached a solution! Please, have your men lower their guns!"

Ernesto Perez looked a little skeptical, but he heeded Alvaro's instructions. He had the loggers and archaeologists lower their weapons. Itotia then lowered Tracey off his shoulder and laid her on the ground. Tracey wasn't sure why he had suddenly done this, but she was very relieved. Alvaro cautiously came over, picked Tracey up, and carried her over near Ernesto, whose face was still full of questions and skepticism.

"What happens now?" he asked Manihuari. "What solution did we reach?"

Alvaro paused as if he thought Ernesto might not like his answer. "They gave us the girl, and I gave them the temple."

"What?" Ernesto whispered fiercely. "You are giving them access to the Huaca Sun Temple?"

Alvaro nodded. "I had to save the girl. Really, I think she is the one who saved us. If she hadn't jumped out of the forest when she did, we may have all been engaged in an unwanted battle right now."

Ernesto looked at Tracey, who was now standing between him and Alvaro. The skepticism left his face as he smiled. "You are right, Alvaro. You did the right thing. Thank you, Tracey. I am glad you're okay."

Tracey smiled and slightly nodded, not feeling the need to be thanked.

"Oh, but there is one more thing, Ernesto," Alvaro said. "I told them they could tie us up while they inspect the temple." Skepticism immediately returned to Ernesto's face.

After an hour or so of racing through the rainforest, Blaine and Arrio had finally caught up with Tenyoa and the other Yora men. They had stopped briefly and regrouped. Then, as a group, they had raced off again. Blaine wasn't sure how close they really were to finding his sister. There was no real way of telling when exactly she was taken. They could be hot on the trail, or they could be hours behind Tracey and her kidnapper.

Blaine's legs churned on. He wondered how much longer he could keep up at this pace. He started daydreaming as he ran. He wished he could be doing this forest chase on motorcycles instead of on foot. Or maybe some kind of hovercraft. Or, better yet, a new vehicle he was inventing in his mind even now—a speed wagon. He and his native friends could flash through the green trees and plants together in this fast and nifty vehicle. Blaine would name it after his local expert. It would be called Arrio Speed Wagon. Suddenly, the design and schematics of Blaine's imaginary vehicle disappeared from his mind as he followed the racing Yora out from the forest into a clearing.

Itotia and the Matsigenka had secured the group extra tightly to trees at the edge of the clearing. Tracey, Alvaro, Ernesto, and every logger and archaeologist now sat hopelessly fastened, watching the indigenous tribe try to open the Huaca Sun Temple.

Just as Arrio had described it the day before, the miniature temple was about the size of a small hut. It was approximately eight feet high and eight feet wide and was shaped somewhat like a pyramid. It was intricately designed, built with hundreds of bricks of different sizes. Tracey remembered Arrio saying to open it would be like solving a puzzle—an extremely difficult puzzle, by the looks of it.

Tracey was tied to a tree with Alvaro and Ernesto, and even now, Ernesto was commenting on how difficult a task he thought opening the temple would be.

"I don't think they will be able to do it," he said, shaking his head. "I know we only found it a couple of days ago, but the archaeologists have been inspecting and trying to unlock it ever since we cleared the jungle away from it. Even with all of their combined expertise, they have no clue about how to open it. The ancient Incans who built this really were quite ingenious."

The big muscular Itotia was taking the lead in trying to unlock the puzzle that was the Huaca Sun Temple. His tribesmen occasionally gestured or made encouraging noises, but he was the only one that was climbing on top of the temple and carefully examining every single brick.

"Tracey, have you heard the legend behind this temple?" Alvaro asked the Sassafras girl.

"Yes, sir, I have," Tracey responded. "There is supposed to be a piece of a magical golden staff inside it. And if it is found by someone with Incan blood, that small piece of staff will guide them to new land."

"That's right," Alvaro replied, impressed by the twelve-year-old's knowledge.

"I love the history and legends we have in Peru," Ernesto interjected. "That is why I have taken such great care of this find. It is why I desire to preserve it by carefully transporting it to a museum. I think it is very appropriate for the tribal people to inspect the temple. After all, it really belongs to them, does it not? I just don't see how they will be able to open it."

Just as Ernesto Perez finished his sentence, Itotia let out a happy grunt. He had found the brick he'd been looking for. It was near the top of the temple. The big Matsigenka man grabbed the stone brick with two hands and twisted it. Then, almost as if he already knew the combination, Itotia moved to another brick and another brick, sliding the second one to the right and pushing the third brick inward. Itotia continued like this, quickly progressing through the complex puzzle brick by brick, sliding, pulling, pushing, and twisting. The eyes of the Matsigenka tribesmen were wide with anticipation and wonder, as were the eyes of everyone else watching Itotia work through the ancient puzzle. Some bricks were high and some were low, but the big man continued to bound all over the temple to the correct spots almost as if the combination was written on his heart.

For a moment, Tracey forgot she was tied up as she wondered, "Will he be able to do it? Will Itotia be able to open this puzzle box? And, if he does, what will be inside?"

By the look of it, Tracey's last question would soon be answered because as Itotia grabbed and rotated a certain last brick, a small door opened up on the front side of the structure. A collective gasp echoed through Itotia's captive audience as he climbed down and faced the open Huaca Sun Temple. The door that had opened was not big enough to walk through. It was only big enough to look through or reach through as it was more like a window than an actual door. Itotia stood there reverently for a second. Then, he slowly reached his big hands and arms inside the opening.

"Could this ancient structure really hold a section of the magic golden staff?" Tracey thought excitedly. A legend she had originally thought sounded pretty far-fetched now seemed like it could possibly be true as Itotia grabbed something that no one could yet see and slowly began pulling it out. His elbows became visible, then his forearms, then finally his hands came out. And, in his hands, he held . . . a pile of dust.

"It's just dust," Ernesto exclaimed. "No golden staff."

But if Ernesto or anyone else was expecting Itotia and the Matsigenka to be discouraged about this find, it was not the case. The big warrior leader held the pile of dust in his hands as if it were a precious newborn baby. He motioned with his head to one of his friends, who immediately opened a leather bag. Itotia carefully poured the mound of dust from his hand into the awaiting bag. When this task was complete, he took the bag from his assistant and held it up in the air. Then, without saying a word, he walked off into the forest, and all of the Matsigenka tribe followed him.

Tracey, Alvaro, Ernesto, the loggers, and the archaeologists sat quiet for a moment. They weren't sure what to make of what they had just witnessed. Was the legend true, or was it not? Though they had all seen the temple puzzle solved and opened, no one really knew.

Eventually, their wonderment gave way to grumbling, as they all began to focus on the discomfort of being tied up to trees. It was anyone's guess how long they would be tied up here or how they could escape. Trying not to think about her sore, chafed wrists and ankles, Tracey let her mind walk through the events of the past couple of days. The study of botany had started with a trip back to Peru, where she and Blaine had reconnected with Arrio, a local Peruvian they had met during their first time here. He had taken them fishing with a bow and arrow. He had shared stories and legends around a campfire. He had begun teaching them about certain rainforest plants, but they weren't yet finished with their four topics of study. They had learned about epiphytes and ferns but not about tropical shrubs or fungi because she had been kidnapped during the night by Itotia. It had all turned out all right because she had found and been reunited with familiar faces. Plus, she had also just witnessed a man open one of the legendary miniature sun temples.

Really, though, what Tracey wanted most was to be reunited

with her brother so they could continue to learn about botany. She didn't have her carabiner, harness, or smartphone. She was tied to a tree. As Tracey thought about her situation and looked out through the forest, her eyes spotted something of interest. It was something Alvaro might be able to tell her about.

"Alvaro, do you see that red stuff that looks like shelves on the bottom of that tree stump over there? Can you tell me more about it?" Tracey asked.

"I sure can." Alvaro nodded. He looked like he was about to start, but then he stopped and laughed. "Didn't we find ourselves in a similar situation last time I was with you?"

Tracey nodded and laughed, too. "Yep. We were caught in Ortiz's nets, and you taught us about blue morpho butterflies."

"That's right," Alvaro remembered. "Well, that red stuff you're talking about is a fungus or mushroom called bracket or shelf fungus. It is common here in the Amazon and can also be white or gray. Bracket fungus grows on dead tree stumps or fallen logs. It has a tough, woody body that resembles shelves. It has spores that are released from tiny pores on the underside of its fruiting body. Interestingly, fungi are neither plants nor animals. They are—"

NAME: Shelf Fungus
DIVISION: Fungi
DISTRIBUTION: Worldwide
HABITAT: In every habitat.

"They are their own group of living things," another voice interrupted Manihuari.

Tracey looked up, and there approaching them was Arrio, Blaine, and a whole group of Yora tribesmen. It was Arrio who had interrupted Alvaro!

With a smile, he kept on going. "The fungi kingdom includes

mushrooms, toadstools, and molds. Fungi lack chlorophyll so they cannot make their own food. Instead, they release enzymes that decompose dead or living organisms near them; then they absorb their nutrients. In the process, they also release chemicals that can be used by other living things again. So, in a way, they are nature's recyclers."

Arrio stopped talking about science and patted Blaine on the shoulder. "We found her, and other than being tied up, she looks safe and in good company."

Blaine smiled and looked relieved, but he also looked surprised to see Alvaro Manihuari and Ernesto Perez. The Yora men pulled out their daggers and began cutting everyone free. As they did, Arrio finished giving information about fungi.

"Fungi are composed of many branching threads that are hidden within their food along with a fruiting body that we can see. This fruiting body produces the spores that eventually grow into new fungi. They reproduce through these spores, which are either dropped from the fungi or shot out from a sac. Either way, they are typically carried by the wind, rain, or insects to grow a new fungus. Overall, the three main differences between fungi and plants are that fungi don't make their own food while plants use photosynthesis to produce their food; fungi are decomposers while plants are producers; and fungi do not have true roots, stems, or leaves while plants have them all."

Alvaro, now free, jumped up and gave his native friend a thankful slap on the arm. "Wow, Arrio, all those books I gave you and all your studying are really paying off. You sound like an expert botanist!"

Alvaro then turned to look at Blaine. "Good to see you again, Blaine."

"Good to see you again, too, Alvaro," the boy responded. "Thanks for taking care of my sister."

"Really, she looked after us. You will have to ask her all about it."

Blaine helped Tracey up as she was cut free. "So now you're trying to run off and have adventures without me, huh," Blaine teased as he gave his sister a one-armed side hug.

"No, no, I wouldn't dream of that," Tracey responded.

"Well, why not?" Blaine retorted.

A playful smirk came across Tracey's face before she answered. "Well, you know why, Blaine. It's because you're such a fun-gi."

Chapter 4: Lost in a Scottish Castle Maze

Roses for the Lady

It was now early afternoon, and the Sassafras twins were sitting in a familiar comfortable place—the Bienvenidos Deck of 'Out on a Limb' guesthouse. This deck served as the common area of the breathtaking treetop guesthouse run by Alvaro Manihuari. At the conclusion of their most recent Amazonian adventure, Alvaro had invited them back for a rest and free cappuccino. Ernesto Perez and their local expert, Arrio, had come along as well.

They all agreed at the end of a long conversation that everything had worked out as well as possible. Tracey had been freed, and no one had been injured during the scary standoff.

Ernesto Perez had pointed out how appropriate it was that a native man with Incan heritage was the one who figured out how to open the Huaca Sun Temple. In the end though, no one really knew what to think about the handful of dust that Itotia had pulled out of the temple's opening. Was it really just dust? Or was it small pieces of the golden staff? Whatever the handful of dust was, they all hoped the Matsigenka found what they needed—a land untouched by humans, one where they could sustain their way of life. Ernesto pledged to follow through on his plan to transport the unearthed miniature sun temple to a Cuzco museum, which both Arrio and Alvaro thought was a good idea.

Eventually, the twins found themselves alone on the Bienvenidos deck. They opened the SCIDAT app and entered the information they had learned along with the pictures they had taken. Then, with tinges of excitement, they hit "SEND" and waited for the next application to open.

"What does LINLOC say, Trace?" Blaine asked his sister as he slipped on his harness.

"Scotland!" Tracey exclaimed. "We are going to a Scottish castle! Longitude is 3° 51' 28.08" W, latitude is 56° 46' 24.96" N. Our local expert's name is Fiona McRay, and our topics for study are roses, hedges, mosses, and peat bogs."

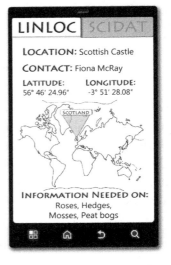

The twins both calibrated their three-ringed carabiners and then waited the several exciting seconds in anticipation for the light-speed zip-lining to begin.

Whoosh! The Sassafras twins zipped off again, one more successful location behind them and one more in front of them. Light, light, light, and stop—where had they landed this time?

Wherever they were, it was cramped. As the Sassafrases' senses returned to normal, they couldn't see much of anything, and the only thing they could feel was cold stone all around them.

"Blaine, where are we?" Tracey grunted as she tried to move around. "Did we calibrate our carabiners to the wrong coordinates? Oh, no, we didn't get stuck in boxes again, did we? Blaine, where are you?"

"I'm below you," Blaine retorted immediately. "And you are standing on my head. Why are you on my head? Get off my head!"

"I'm sorry, I didn't know."

"Your foot is right on my forehead!"

"Calm down! I'm trying to move." Tracey squirmed around, trying to reposition her body.

Blaine pushed up to try to get his sister's foot off his face. Fortunately, the two managed to wriggle free from their tight spot. Unfortunately, they were now falling.

This time, the fall didn't last very long, but they landed in an awkward, but painless, thud on a small pile of logs. Blaine looked at Tracey, and Tracey looked at Blaine. They were both covered in black soot.

"I think we landed in a fireplace," snickered the Sassafras girl.

Blaine was about to make a joke about Santa Claus when suddenly a set of doors burst open, spilling light into the room that lay in front of the twins. They clearly saw that they had indeed landed in a fireplace. Thankfully, there was a screen protecting them from being spotted by the three people who entered the room. They were led by a tall gangly man with a few wisps of hair left on his head. He had long facial features (especially his nose) and dark eyes, and he was dressed in a tuxedo with tails. The man barked orders in a Scottish accent to two women who were following him. One woman was thin and dressed in a maid's uniform, while the other was plump and dressed like a baker.

"It is almost nine o'clock at night! That means the guest of honor will be here any minute! Need I remind you two that Lady Dockerty wants this room looking perfect? You must get on with the final preparations right now!"

With that, the man in the tuxedo turned in a huff and exited the room. As soon as he was gone, the two women laughed.

"Dunmore is such a hissy-fitter, isn't he now, Osla?" the baker asked the maid.

"Aye, that he is, Rona," the maid agreed. "I mean, just look at this room. It's already been perfectly prepared. What else does he think needs to be done?"

"True, but just to make our butler happy, I'll go down to the kitchen to grab a few more pastries to bring up. Maybe you can dust the drapes or something," Rona suggested.

Osla the maid smiled and laughed. "Rona, I dusted the drapes not half an hour ago, but I guess it can't hurt to do it again. Maybe I can also light a few candles?"

"Sounds good, dearie," replied Rona the baker as she left to get more of the pastries.

As Osla dusted the drapes and lit some candles, the Sassafras twins looked around the big room through the gaps in the fireplace screen. It looked to be a grand sitting room of some kind with a high ceiling, tall windows, several luxurious couches, a perfectly polished table, and exquisite rugs and tapestries. The room was accented with expensive-looking vases full of beautiful flowers. A grand wooden table, which was in the center of the room, was covered with yet more flowers. The table also had candles and a variety of fine china overflowing with pastries.

"Wow," Blaine whispered. "This is quite the spread! Let's go dig in!"

Tracey grabbed her brother's arm before he could bolt out of the fireplace. "Blaine!" she murmured quietly but directly. "We can't go trouncing out into that room. Just look at us! We are covered in soot!"

"But it looks so good!" Blaine moaned.

"We are not going out looking like we do in a place this nice. We have to stay put," Tracey commanded.

Blaine sighed in disagreement, but he stayed in his spot nonetheless.

After a few minutes, Rona returned with two more dishes of pastries, and right on her heels was Dunmore. The butler was ushering in three new people. "I'll never know why Lady Dockerty invited all you people to this occasion, but she did, so wait in here until the Lady and her brother arrive and then finally the guest of honor. Please try not to break anything."

Again, the butler turned and left in a hurry as if the world were about to end. There were now three new characters in the room, and they were indeed characters. The first was a fairly big man wearing all denim with grass stains on his knees. He sported a wide-brimmed sun hat and held a pair of lopping shears. The second was a young woman who was the most normal looking of the bunch, except for the fact that she had the thickest, wildest, reddest hair the twins had ever seen. It made Uncle Cecil's red hair look tame in comparison. She also had a face full of pretty freckles. Finally, there was a bearded man wearing knee-high argyle socks and a kilt. He was holding a set of bagpipes.

"Well, lookie here, Osla." Rona chuckled. "Look what the cat dragged in. The piper, the gardener, and that wee bonny lass."

"My name is Fiona, and I'm a botanist, not a wee bonny lass," the redheaded woman asserted.

"So, this is our new local expert," thought the twins.

"No need to get lippy, dearie," Rona shot back, not intimidated by the fiery redhead. "Just making conversation."

"By the looks of it," the gardener spoke up, "Lady Dockerty invited us this evening to class the place up a bit."

"Oh, Angus," Rona said. "Really? Class up? You smell like dirt, and you brought your shears inside again. You're going to kill poor Osla. She has to clean up after you, you know."

Osla just smiled at Angus the gardener like she didn't mind. Then, without warning, the kilted man began blowing into his bagpipe, causing loud music to fill the room.

"Leif! Leif!" Rona cried as she ran over and pulled the instrument from the piper's mouth. "You're going to give Dunmore a heart attack! You know you're not supposed to play the pipes unless asked to do so by Lady Dockerty herself!"

Leif the piper's emotionless face remained unchanged as he let the sound of the pipes slowly fade to silence.

Minutes later, Dunmore the butler was back. This time, however, he was not frantic. Instead, he was calm and composed. This time, he brought with him the lady of the house. Dunmore stopped by the door, stood with perfect posture, and announced, "Lady Dockerty and her esteemed brother, Sir Kentalot."

Angus and Leif bowed slightly. Rona and Osla curtsied. Fiona just stood with a smile on her face that could have been genuine or phony. It was hard to tell. Lady Dockerty was dressed like royalty in a sophisticated dress complete with white elbow-length gloves. She looked to be sixty to seventy years old, but she carried herself in a strong and graceful way. She had sharp facial features and a head of perfectly styled gray hair. Her brother, also in his sixties or seventies, was wearing what looked to be a naval officer's uniform covered with medals and patches. He had a gray head of hair and a big bushy gray mustache to match.

"My, what a lovely staff you have, dear sister," Kentalot bellowed in a deep but cheerful voice, referring to those already in the room. "They remind me of the great staff that I had supporting me during the war. It was the winter of nineteen hundred and—"

"Oh, enough about your war." Lady Dockerty interrupted her brother. "Tonight is not about you! It's about our guest of honor. And he is bound to be here any minute. Isn't that right, Dunmore?"

The butler nodded. "Yes, madame. He is slated to arrive at nine o'clock, and it is just now five minutes 'til."

"Well, excellent, then." Lady Dockerty smiled. "All of you must now consider yourselves off-duty. Make yourselves comfortable and get ready to welcome my son back home."

Dunmore, seemingly the busiest person in the house, left again. The twins presumed it was to go and wait for Lady Dockerty's son, the guest of honor. Everyone else took seats on various couches in the room, except for Osla, who seemed nervous. She told Lady Dockerty how lovely she looked and then continued to bustle about, using her feather duster to dust things that didn't look dirty at all.

Blaine was uncomfortable in the fireplace. He felt his legs beginning to cramp and fall asleep. If that gardener could be a part of the party with his grass stains and lopping shears, why couldn't he be a part of the party with soot stains and a harness? He was about to share his sentiments with Tracey when Dunmore's voice suddenly interrupted his thought.

"Presenting this evening's guest of honor . . . Mr. Miles Dockerty."

"Miles Dockerty!" Blaine hissed way too loudly.

Tracey reached up and grabbed her brother's mouth to shush him. Blaine pulled Tracey's hand away. "Tracey! It's Miles Dockerty! Don't you remember who he is?"

Tracey had a look on her face like either she had forgotten or that she didn't care and just wanted her brother to stop talking.

"He was one of the judges at TOBA in Sydney," Blaine whispered emphatically.

Realization hit Tracey's face. Blaine was right. The girl looked out from the fireplace at the guest of honor. It was Miles Dockerty. They hadn't met him, but they had seen him when they were in Sydney, Australia, studying the respiratory system. He was one of three judges who had decided the winner of the "Take Our Breath Away" (TOBA) talent competition.

The twins had noticed then and they were noticing again that Miles Dockerty wasn't big on style or personality. He had slicked-down hair and dated glasses, and he was wearing a sweater vest. He wasn't frowning, but he wasn't smiling, either. Regardless, he stood with his hands behind his back, not looking thrilled to be at this party his mother was putting on for him, and he looked a little wet. "It must be raining outside," the twins guessed.

Lady Dockerty stood up and smiled. "Miles! Welcome home, son!" She then turned to the hodgepodge group of characters in the room. "Everyone stand up and welcome Miles home. He has just finished another successful season as a judge on the internationally televised talent show, 'Take Our Breath Away.'"

Everyone stood up and did their own version of welcoming the night's guest of honor. Angus and Leif bowed slightly again, Rona curtsied, Sir Kentalot bellowed salutations, and Osla and Fiona stood still and smiled, although, Osla's smile looked more sincere than Fiona's. Dunmore grabbed the two large wooden doors to the room and closed them to officially start the celebration. Inside the grand sitting room, all was bright and festive despite the rain that began to patter on the windows. The only exception was the dark and dirty fireplace where the twins were hidden.

With his hands still behind his back, Miles stepped toward the center of the room to the table. "I have a gift for you, Mother—a dozen roses."

The talent judge carefully placed a rectangular glass box full of roses, which he had been hiding behind him, in the center of the big wooden table. Everyone in the room gasped because the "roses"

Miles had just put down were not real flowers at all. They were actually sparkling red jewels balancing on the end of clear, crystal stems.

Lady Dockerty covered her open mouth with her hands for a second and then exclaimed, "Oh, Miles, what an exquisite gift!"

"I got these for you while in Sydney, Mother. Each one is a fourteen-carat ruby, set in a stem of pure crystal. I thought they would go nicely with the vast collection of flowers and plants you have here at the castle."

"Oh, my dear son, they will fit very nicely here, except for the fact that they will stand out as the most beautiful. Miles, you really shouldn't have!" Lady Dockerty gave her son a hug, as Miles just stood statue-like.

Everyone else in the room stood smiling and looking at the roses. It was obvious none of them had ever seen anything like this before—twelve beautiful shining rubies set on the end of perfectly designed sparkling crystal stems. The eyes of the guests revealed the joy and wonder they felt as they gazed upon the gemstone flowers. Their eyes also exposed how their minds were rapidly calculating just how much the dozen roses might be worth.

"Miles, my lad," Sir Kentalot grunted. "These roses are spectacular. I haven't seen beauty like this since that fateful night during the war. It was the winter of nineteen hundred and—"

The naval officer's sentence was interrupted by a loud crack of thunder. Suddenly, the well-lit room went dark. The Sassafrases' senses shot into high gear as a series of curious noises sounded in the darkness. First, there was a grunt. Then, there was a brief sound of air being released from bagpipes. Next, there was the sound of shattering glass, proceeded by a high-pitched scream. There was another grunt as footsteps could be heard scurrying around. Then, the creaking sound of a door opening. And, finally, came several minutes of eerie silence.

Blaine was about to whisper something to Tracey, who was to his right, when he suddenly felt the presence of another person crouching to his left. All at once, the lights in the room came back on, revealing a very different scene from the one before they had turned off.

Miles, Sir Kentalot, Rona, and Osla were missing. Dunmore stood at a breaker box on the far wall. Lady Dockerty sat looking shocked on a couch, as Angus stood over the unconscious body of Leif. Most shocking of all to the twins was that Fiona McRay now sat crouched next to them, silently puffing for air and sharing the twins' pile of soot-covered logs. Fiona looked over at the twins with questioning eyes, much like their own. However, none of them said a word.

The silence was broken by a cry from the gardener as he screamed, "Leif the piper is down! I repeat, Leif the piper is down!"

Dunmore closed the breaker box and rushed over to where Leif had fallen unconscious. The butler kneeled over the piper and gave him a good slap in the face. Leif shot up, immediately grabbed his bagpipe, and started playing it. Lady Dockerty, meanwhile, stood up and walked over to the wooden table. Everything on the table had previously been neatly placed. Now the surface looked like the scene of a rough-and-tumble brawl. Candles were knocked over, pastries lay strewn about, and in the center of it all lay a pair of lopping shears next to sharp pieces of shattered glass.

More important than what was there though was what *was not* there. "The dozen roses! They are missing!" Lady Dockerty gasped.

Dunmore implored Leif to stop playing the bagpipes as he stood staring at the table with his stately employer. He looked like he was about to say something when the four missing employees suddenly barged back into the room. They all ran straight to the table, joining the butler and the Lady in gazing at the spot where the roses had been. The gardener and the piper joined the group as well.

All eight stood in silence for several frozen minutes. The Sassafrases assumed that everyone was trying to figure out what had happened to the jewels.

Almost all of the eyes remained glued to the wrecked table. All but Dunmore's eyes, which scanned the faces of his seven companions. It was obvious he was looking for guilt, and he stopped when he got to Angus.

"It was you!" The butler pointed to the gardener.

Angus immediately looked defensive. "Me?" he asked as if the idea was preposterous. "Why do you say that?"

"Because it's your lopping shears that are there on the table. They must be there because you used them to break the glass box. Which means, you sir, stole the rubies!"

"Someone yanked them out of my hand," the gardener defended himself. "Besides, I didn't even leave the room. What about these four?" Angus gestured toward Miles, Sir Kentalot, Rona, and Osla. "They all left the room. They had plenty of time to take the jewels and hide them somewhere in the castle!"

Dunmore turned his gaze from Angus and stared accusingly at the four that had left the room.

"Don't look at me!" Miles squeaked out. "Why would I steal a gift I just gave away?"

"My honor is irreproachable," Kentalot bellowed.

"I ran from the room because I was scared, Dunmore!" Rona screeched.

"Me, too!" added Osla, smiling sheepishly.

Dunmore looked at the last two standing at the table. "Well, obviously it wasn't the honorable Lady Dockerty, and it couldn't have been Leif because he was knocked unconscious. Who does that leave?"

"It leaves you," Angus accused. "Tell me, Dunmore, why

aren't you under suspicion, too?"

"Because it simply was not me," Dunmore stated with an air of authority.

"Hey! Where is that new wee lass?" Rona interjected.

"Yes, where is that new botanist?" Osla asked.

"I had completely forgotten about Miss McRay!" Dunmore said as he slammed his right fist into his left palm. "She must be the thief! Why else would she run and not come back? Hurry, everyone, spread out and search the castle. We must not let her get away!"

The party of eight quickly dispersed, bent on their mission to find the perpetrator, leaving the grand sitting room empty. Blaine and Tracey slowly turned and looked at the redheaded botanist next to them in the fireplace. Was their local expert a jewel thief?

Blaine couldn't bear the suspense, so he just came out and asked her, "Did you steal the roses?"

Fiona looked at him with an annoyed face as she responded, "No. I did not. Did you?"

"Me?" Blaine laughed nervously at the accusatory question before answering, "No way!"

"Then why are you hiding in the fireplace?"

"Why . . . the fireplace . . . you ask . . . It's because . . . the, uh . . . castle is . . . it's because . . ."

"We came to this castle to learn about the flowers and the other plants they have here, not to steal a dozen roses." Tracey helped her stammering brother.

"Well, why didn't you say so?" Fiona chuckled. "I am the botanist at this castle. Even though I have worked here only a couple of weeks, I can tell you everything there is to know about the plant species we have here."

Fiona paused and looked sternly at Tracey before she

continued. "But, please, little lassie, don't call those rubies, or whatever they were, roses. It's an insult to the flower. Roses are much more beautiful than those fancy, over-priced pieces of stone. There are over one hundred varieties of roses, most of which are in shades of yellow and red. We have eighty-seven of those varieties here on the grounds of Dockerty castle. We frequently use them as cut flowers, but roses are also used in perfumes and occasionally in food and tea. These flowers have a tough, woody stem that often has thorns."

"Roses can climb over trellises, ramble along the ground, or grow as shrubs." McRay continued. "Their flowers are very showy and scented to attract insects for pollination. The sepals fall off as the petals of the flower open."

"What are sepals?" Blaine asked, getting a little more serious.

"Flowers have several key parts, all of which form around a central axis. The sepals form a protective layer around the petals while the flower is budding. The job of the petals is to attract insects for pollination. In the center of the flower are the reproductive structures—the pistil, which contains the female reproductive structures known as the ovary and the ovules, and the stamen, which contains male reproductive structures known as the anther and pollen."

As Fiona McRay kept talking, she got louder and more excited. The twins wondered if that was a good idea, considering she was currently being sought as a jewel thief. Then, she surprised Blaine and Tracey by actually crawling out of the fireplace, pulling the twins with her as she walked over to a vase that was on an indented shelf in the wall, with light shining down to illuminate the flower that it held.

"This is a species of red rose," the botanist said. "See? Isn't it much prettier than those silly rubies with those ridiculous crystal stems? It is an amazing angiosperm, which another way of saying 'flowering plant.' These plants produce seeds that are protected by a fruit."

The twins nodded instantly in agreement since they were nervous that someone was going to hear them and burst back into the room. They grabbed their phones and quickly snapped a photo of the rose in the vase.

Fiona was about to say something else when she was suddenly interrupted by a crashing sound on the other side of the room. All three looked toward where the sound had come, but they saw nothing. The redheaded botanist took that as their cue to leave.

"Let's get out of here, before those crazies come back in here and find us," she suggested. "I know several good hiding places!" With that, Fiona McRay bolted out of the room with the Sassafras twins following right behind her.

A Maze of Boxwoods

If he had landed just a few minutes sooner and been just a little bit quieter, he potentially could have had the twins' phones and the dozen rubies. But, as he stood now, he had neither. He had just clumsily knocked over a fancy flowerpot, causing the crashing noise. The sound had startled the twins and their new local expert. They had all swiftly exited the room. As they had scampered out, they had passed right in front of him as if they had not even seen him! Wearing the Dark Cape suit would surely pay off sooner or later. He carefully picked up the flowerpot, and then he took off into the hallway behind the three.

"This castle is huge," thought the twins as they raced behind Fiona McRay. They had started in a wide hall with a high arched ceiling. Then, they had ducked off through a doorway to the right, entering a ballroom of some kind. They were now hurrying through adjoining rooms connected by ornate wooden doors. All at once, Fiona stopped, causing Tracey to run into the back of her, and Blaine into the back of Tracey.

The botanist turned around, her freckles crinkled up in a perturbed scowl. She pressed her index finger to her lips, instructing the twins to be quiet. The botanist had stopped because she had heard something, and now the twins heard it, too.

It was the voice of Rona the baker. "Dunmore! I think I saw her! She ran into the ballroom, and she had two other people with her! Two wee people!"

"Very good!" The three now heard Dunmore's voice echoing through the castle. "Find her and stop her if you can. Lady Dockerty is now fetching Duke and Earl!"

"Duke and Earl?" the baker asked.

"Yes, Duke and Earl," Dunmore shouted. "The dogs, Rona, the dogs!"

"Oh, yes!" Rona acknowledged. "Duke and Earl, the dogs. They will be helpful in aiding our search, won't they now?"

"Duke and Earl?" Tracey whispered in fear.

"The dogs?" Blaine screeched, sounding scared.

Fiona looked more agitated than scared as she whispered emphatically, "Duke and Earl won't be a problem, and neither will Dunmore, Rona, or anybody else. Just follow me."

Fiona tiptoed out of the room into the hallway. They saw Osla, Miles, and Sir Kentalot going from the hallway into one of the rooms they had just left. Fiona scampered down the hallway a few yards and then led the twins into yet another room. As they entered that room, they caught the back of Rona exiting the room for the hallway, using another door. McRay and the Sassafrases waited perfectly still for a few long seconds and then stepped back into the hallway again.

The game of hide and seek continued on like this for several minutes as the three sneaked in and out of rooms while the baker, the maid, the butler, the naval officer, and the talent show judge pursued them. The crisscrossing got so confusing it was almost impossible to know who was doing the hiding and who was doing the seeking. Then, to make it even more confusing, Lady Dockerty with her dogs and Leif with his pipes joined the search party. The dogs barked, Leif's pipes played, and the chase continued. At the height of the confusion, Fiona and the twins slipped out of a room into the hallway and then up a stairway at the end of the hall.

Their exit went unnoticed by their pursuers, who continued on with the carousel of confusion. McRay skipped the first two steps and jumped up to the third step as she led the way up the wide circular staircase. She was fast, but so were the twins. As they climbed, they saw more beautiful flowers on shelves along the way, as well as large intricately woven tapestries, just like there were in the grand sitting room. As she ran, Fiona reached and pulled off several of the decorative ropes that ran along the tops and sides of the tapestries. The twins weren't sure why, but they didn't want to ask, either.

McRay raced past the second floor up to the third. Upon reaching the third floor, the three stopped to catch their breath. There were no lights on, and all was dark and silent, except the faint sound of . . . barking.

"Duke and Earl," Blaine exclaimed.

Tracey added, "They are coming up the stairs!"

Fiona looked to her right and left. She paused for a moment and then took off to the left, deciding that direction was the best way to go. The barking got louder as the dogs got closer. The three raced down the dark upstairs hallway until they hit what looked like a dead end.

"Oh, no!" Tracey yelped. "We're trapped."

"We're not trapped, lassie," Fiona retorted. "There is a door here. It's just locked. Give me a minute, and I'll be able to pick it."

"We may not have a minute," Blaine suggested with a worry-filled voice. "It sounds like Duke and Earl are about to reach the third floor!"

"Oh, calm down, laddie," the botanist assured as she pulled out a hairpin that had been hidden somewhere in her wild hair. "I've done this plenty of times."

Fiona sounded confident, but the twins weren't so sure. They looked down the hall and saw that the dogs had reached the third floor. The Sassafrases could now see that Duke and Earl were full-sized Scottish terriers. They were not especially large or imposing animals, but any dog that is running straight toward a person while barking ferociously looks daunting.

Fiona stuck the hairpin into the lock and began trying to jimmy the door open. The twins, who had never picked a lock, thought it looked like an impossible task, especially considering the fact that this lock was in a dark hallway and mad terriers were coming at them. McRay wiggled the hairpin, while Duke and Earl continued to spring toward them.

"By cracky," Fiona exclaimed. "This is a tough one, but I think I almost have it."

"Almost" was not a word the twins were hoping to hear because Duke and Earl were quickly approaching.

Now Duke and Earl were only feet away from them, but suddenly one of the strangest things the twins had ever beheld happened. The dogs stopped just short of them and started acting as if they were attacking an invisible man. Right then, Blaine and Tracey heard the door open. McRay let out a little shout of victory.

She slipped into the room and quickly yanked the twins in behind her. Immediately, she closed the big wooden door and locked it.

People couldn't see him or sense his presence, but evidently, the dogs could. He continued to silently and invisibly panicked as the dogs called Duke and Earl nipped at him. He had been just mere inches away from nabbing the twins' smartphones, but then these crazy dogs interfered. Being invisible, he thought they would have run right past him and attacked the three visible people, but they had some strange canine sense that made them go for him instead. What could he do now? How could he escape Duke and Earl?

"Maybe if I suddenly switch to visible," he thought, "it will scare these dogs long enough for me to get away!"

He tugged the vanish string and immediately he was standing in the hallway as the Dark Cape. Duke and Earl dropped back and stopped barking for just a second. It had worked!

He immediately took off down the hallway, back toward the circular staircase. After only a moment's pause, the hyper Scottish terriers were back on the job, barking and chasing the stranger. As he ran toward the light at the end of the long hallway, he saw something that made him gulp. The lady, the naval officer, the talent connoisseur, the butler, the maid, the baker, the gardener, and the piper had all reached the top of the staircase as a group. This was not good.

The moment the group saw him rushing toward them in the Dark Cape suit, all eight froze in fear. Just before he reached them, he pulled the vanish string again and disappeared right in front of their wide eyes. He rushed invisibly past them and then down the stairs. As he wound downward, he could hear Duke and Earl still barking and giving chase. But they would not catch him because he was about to zip away from this place. He would have to wait until the next location to steal the twins' phones!

Blaine and Tracey now found themselves hanging precariously outside a third-floor window as lightning flashed, thunder sounded, and rain pelted them. Fiona had tied together all of the decorative ropes from the tapestries. She had then secured the ropes to a post in the room and tossed them out of the window. The three of them were now climbing down the length of the ropes from the third floor, hopefully to the ground. The agile botanist, who was the furthest down on the rope, was the first to reach the ground. The twins then hurriedly followed the redhead to the ground and out of the rain under the shelter of a nearby gazebo. The three of them just stood there quietly dripping for several minutes.

Blaine eventually broke the silence. "Well, I guess we got away." He tried to sound positive as they stood in the rain.

"We didn't get away, laddie," Fiona replied. "We are out of the castle, yes, but now we are stuck amongst the hedges. This gazebo stands at the beginning of the largest hedge maze in all of Scotland."

"Hedge maze? What's a hedge maze?" Tracey asked.

"It's a confusingly intricate network of passages and walkways with hedges for walls," answered McRay. "Most hedge mazes that I have seen are complicated, but they have multiple entry and exit points, but not this one. Not the Dockerty maze. It has only one entrance and one exit. The entrance to the maze is the back door of the castle, and the exit is a rounded wooden door somewhere on the other side of all these walls and walls of hedges. There are no other ways in or out."

Fiona stopped and shook her head as if thinking about the

challenge that the maze presented. "Lady Dockerty loves her flowers and has many displayed inside, as you saw. She has lush gardens rolling across the front and sides of her castle, but here behind her castle is her pride and joy—the Dockerty hedge maze."

"These hedges are formed from shrubs grown close together and trimmed to form a barrier around a field or garden plot. In England and Scotland, hedges have become an important part of the ecosystem, housing many small animals, birds, and insects. The hedges are usually formed by slow-growing evergreen shrubs, such as holly or box plant. They are trimmed, so that growth is promoted around the base of the plant.

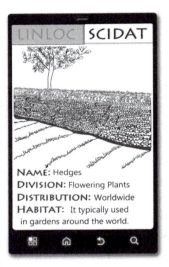

NAME: Hedges
DIVISION: Flowering Plants
DISTRIBUTION: Worldwide
HABITAT: It typically used in gardens around the world.

"The box plant is an evergreen shrub or small tree that can grow two to twelve feet tall. It has small yellowish-green flowers. Although it is normally grown as a hedge or topiary, it can also be found in the wild in Asia, Europe, Africa, and the Americas. The wood from the box plant can be used for small wood carvings and instrument pieces. In fact, it was used to make bagpipes before everyone switched to using imported woods instead."

Fiona paused and looked out beyond the gazebo at the hedges she had just spoken about. "The lines of the Dockerty hedge maze are made up of box plants. Its walls stand over ten feet tall, and, like I said, it is the largest in all of Scotland. It is basically an impossible, unsolvable puzzle. I have never been able to make my way successfully through it."

"But if we want to get away, we have to go through it, right?" asked Blaine.

Fiona looked at the boy and nodded. "Right."

Chapter 5: The Mystery of the Stolen Roses

Mossy Mysteries

Blaine and Tracey guessed it had to be well after midnight. They had been plodding through the Dockerty hedge maze for hours now, and just as Fiona McRay had said, it was big, intricate, and confusing. Both of the Sassafras twins were pretty good at doing mazes and puzzles, but *doing* a puzzle and *being in* a puzzle were two completely different things. They had taken rights and lefts, lefts and rights, gone straight, gone backwards, but the only direction they seemed to truly accomplish was circular. The continually pouring rain wasn't helping matters either, and though there were lampposts throughout the maze, it was still very dark.

At first, the idea of a hedge maze had sounded fun, and maybe it would be fun under different circumstances. However, the Sassafrases were getting tired and discouraged and a little claustrophobic. They rounded a curve in the hedges and came upon a cement bench. Fiona, who was leading the troupe, was obviously tired as well. She saw the bench and took the opportunity to sit down. The twins joined her. They hadn't heard the sound of anyone chasing them since they had left the castle, so they figured they could afford a rest.

Fiona took the break as an opportunity to share about pollination. She began by saying, "So much of a flower's existence revolves around pollination. In pollination, the pollen from a flower lands on the pistil of the same species of plant. Then the pollen sprouts a tube down to the ovule, where the pollen and the ovule meet and join. This eventually forms a seed. A flower can be pollinated by self-pollination, which means the plant pollinates itself, or it can be pollinated by cross-pollination, which means that the wind or animals carry the pollen between plants.

"All flowering plants have roots, leaves, stems, flowers, and seeds." McRay continued. "As botanists, we classify each one according to the number of its seed leaves, which we call cotyledons. Monocots are flowering plants that sprout with a single seed leaf, such as the tulip, iris, or orchid. These plants have narrow leaves, and their flower petals occur in multiples of three. Dicots are flowering plants with two seed leaves, such as roses and different trees. These plants have a central stem that everything grows from. In addition, the flower petals of dicots usually occur in multiples of four or five."

A hush fell over the twins as they thought about what Fiona had just told them. They watched as the rain changed from a downpour to a soft drizzle. The slow-moving droplets floated down beautifully under the dim light cast by a nearby lamppost. It was a hypnotic sight that gently lulled the trio to sleep.

Hours later, Blaine and Tracey awoke with a start at the sound of Fiona's words. "I didn't steal the jewels, so why am I running?"

"Oh, it's already morning." Blaine yawned as he looked around at the light-filled maze. After rubbing his eyes and stretching a bit, he pulled out his smartphone to take a picture of the hedges around him in the morning light.

"And we're dry." Tracey stretched and smiled before taking her own picture.

"If it wasn't me, then who was it?" Fiona pondered, ignoring the twins. "I'm assuming it wasn't you because you were in the fireplace before me, and you were never anywhere near the rubies. It definitely wasn't Lady Dockerty—it is just not her style. Besides, you can't really steal something that is already yours to begin with. The seven others, however, all have motive."

"Even Miles?" Tracey asked. "Why would he steal a gift he'd just given?"

"Because he's a selfish grump," Fiona replied. "He knows that anything his mother has now will one day be his, but maybe he got impatient. Maybe after he gave his mother the rubies, he immediately regretted it. Maybe he wanted them right back. Besides, he didn't seem to really be enjoying the party, now, did he? What better way to kill a party than a quick smash and grab?"

"That seems a little farfetched, doesn't it?" Blaine asked.

"Maybe, but stranger things have happened," the botanist stated.

"What about everyone else?" Tracey asked. "What would their motives be for robbery?"

"Let's run it down from the top—keeping in mind that there was enough time in the darkness for each one of the lot to steal the rubies and hide them," Fiona replied as she leaned forward on the bench. "First, there is Sir Kentalot, Lady Dockerty's brother. He doesn't seem likely to be the thief. However, I have heard that war story of his, and it brings up some interesting points. Back on a winter's night sometime during the early nineteen hundreds, Kentalot claims that he happened upon a find that would change the tide of the war. The story goes that he found a chest full of rubies buried in the snow, which he used to pay the hired mercenaries on the other side to stop fighting. If that tale is true, then rubies would be very sentimental to him, now, wouldn't they? Could they be sentimental enough to cause him to steal his sister's rubies? Maybe, maybe not.

"Next is Dunmore the butler." Fiona continued. "Did you notice how right after the lights came on he was standing by the breaker box? Was that because he was resetting the power, or was he hiding the stolen rubies inside? Did you observe how it was he who immediately started accusing everyone else? Was that because he was trying to shift the blame off himself to someone different? The man knows the ins and outs of the castle better than anyone else. If anybody could effectively hide the rubies, my money would be on him.

"The next person we need to consider is Angus the gardener. Remember what happened after the lights went out. There was a specific series of sounds—a grunt, bagpipes, shattering glass, a scream, another grunt, footsteps, a creaking door, and finally silence. Could it be that most of those sounds were caused by the gardener? Imagine with me, if you can, that Angus was disoriented in the dark but badly wanted those rubies. He ran into Leif, causing the piper to grunt and fall down on his pipes. Angus then swung his lopping shears, broke the rectangular glass box that the ruby roses were encased in, and freed his prize. The next scream and grunt could have come from anyone as Angus pushed his way through the dark to the door with heavy footsteps. He then opened the door, causing it to creak. He hid the rubies, and quickly returned to his spot standing over the unconscious piper, where he silently waited for the lights to come back on. It might also interest you to know that just yesterday Lady Dockerty told Angus she was going to add to his workload, but pay him the same. After this news, Angus confided in me that he was going to make her sorry for not raising his pay.

"Up next is Rona the baker. She is one of the four who suspiciously left the room and didn't return until well after the lights came back on. This gave her enough time to hide the stolen rubies just about anywhere. But something you may be even more interested in is that she has become infamous here at Dockerty Castle. I'm sure you have heard the term 'baker's dozen.' As you know, it came about when generous bakers of the days past would add one extra to a dozen of whatever it was they were making. Not so with Rona here at Dockerty Castle! When people say 'the baker's dozen' here, it means there are only eleven because Rona is known for taking one away to keep for herself. If the plump baker is so free to swipe food, would she not also be free to swipe jewels?

"Then there is Osla the maid. He words and demeanor are nice enough, I guess, but she is a bit flirtatious, don't you think? Her words are always filled with bits of flattery. She is always over-complimenting the lady of the castle and can also be seen batting her pretty eyes and smiling her sweet smiles at every man who crosses her path. Would you believe it if I told you that just last week, as I was in the lady's chambers pruning some flowers, I caught Osla in Lady Dockerty's wardrobe? She was trying on one of the lady's dresses and pinning on a pair of the lady's earrings. She said she was just playing make believe, but that's a bit suspicious, is it not?

"Last but not least, we have Leif the piper. It would initially appear that he should not be under suspicion because he was knocked unconscious during the robbery. But could it be that he was just pretending to be unco—" Fiona suddenly stopped her long deliberation and sat up straight.

"Do the two of you hear that?" she asked.

The Sassafras twins, who had been listening with fascination to McRay, broke from their reverie and perked up their ears. The sound they heard did not make them happy.

"Duke and Earl!" Blaine announced with alarm. "They're back on the chase!"

"That they are. I must have woken them up, too," Fiona affirmed, standing up from the cement bench. Her resolve not to run looked a little shaky. "I guess if we want to escape, it's time for us to figure out this hedge maze now, isn't it?"

THE SASSAFRAS SCIENCE ADVENTURES

The two feisty Scottish terriers sounded far away, but there was no telling how long it would take to get out of here as the three of them hadn't exactly proven themselves capable of making any headway with the maze. Blaine and Tracey hopped off the bench and joined Fiona, who was already racing around the curve, which angled out into a long straightaway. At the end of the straightaway, the three had the choice to go left or right, and they chose to go right. They then took another straight path, rounded a corner to the left, and then hooked a U-shaped right. They were about to continue racing, but instead they ran right into a cement bench next to a lamppost.

"Is that the bench we were just sitting on?" Tracey asked in frustrated confusion.

"I don't know," Blaine responded. "But it sounds like Duke and Earl are getting closer."

The three raced on, though they had no idea which direction they were really going. They could be running away from the dogs, or they could be running straight toward them. They could be getting closer to the wooden door that exited the maze, or they could be heading right back toward the back of the castle. They didn't know. The labyrinth of tall green hedges was proving to be just as difficult in the daylight as it had been in the darkness of night. How were they ever going to find their way out?

The three maze racers came around a corner and stumbled out into a rare open area. It was circular, it had at least a dozen passages leading to and from it, and it had a beautiful rose garden right in the center.

"Well, slap me daft," McRay said with a rare smile. "I know where we are! This is the rose garden that is at the very center of the hedge maze!"

"So, does that mean we are halfway out?" Tracey asked.

"Maybe, lassie, but . . ." Fiona looked like she was about to explain her "maybe," when, suddenly, two barking things burst from an opening into the garden.

"Duke and Earl!" Blaine shouted.

The botanist and the twins turned and shot toward an exit, only they all shot in different directions! Blaine, Tracey, and Fiona now each unknowingly traveled through the hedge maze alone. Duke's and Earl's inexhaustible barking continued as they maintained their unrelenting chase. Blaine looked to his left and then his right and then glanced behind him.

"Oh, great," he thought. "I lost Tracey and Fiona!" Blaine scampered forward through the tall green hedges. He was discouraged that he had lost his companions, but he was determined not to give up. After hours of being in this maze, Blaine felt like he was beginning to figure it out.

Tracey had originally thought she had exited the same way as her brother and Fiona, but she then quickly realized she was alone. However, she was soon joined by two companions, only the companions were not Blaine and her local expert. The companions were canine.

Tracey wanted to scream. Instead, she let determination take over. She would decode this hedge maze and keep these pesky terriers from catching her. Tracey saw an opening in the hedges to her left. She passed it by because she recognized it as one she had already gone through. The two dogs were getting closer, as she saw just up ahead to the right—another opening. Tracey knew it was the one she should take.

She reached it before the dogs reached her. She skipped through it and found herself racing down a zigzag patterned passageway. This was new, and new was good. Duke and Earl were still behind her, but at least she was no longer going in circles.

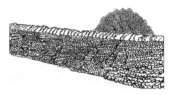

Blaine had slowed down his pace a bit to make sure he was making the right choices about which passageways to take. He could still hear Duke and Earl barking. They sounded nearby, but close or not, the boy wasn't going to let the presence of the dogs deter him from figuring out this giant green puzzle. Blaine took a right, which he was positive he had not taken before, and found himself in a zigzag passageway. Yes, he had definitely not gone this way before.

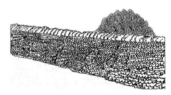

At the end of the first zigzag passageway was a U-turn, which Tracey had taken without pause. She now found herself running back the way she had just come, in an adjacent zigzag passageway. She still felt as though she was going the right way, but with every zig and every zag, she could tell the terriers were gaining on her. Tracey raced past a zig and prepared to zag when she looked back and saw a smaller-than-normal opening in the hedge wall. The girl skidded to a stop, retraced a couple of steps, and plunged through the small opening. Just as she did, barking fur sped past the hole. Duke and Earl had overlooked the opening! She had thrown them off her trail!

Blaine went through every zig and every zag with the fluidity of a professional football running back. He continued to sense that he was going the right way, but he was also sensing that he may be approaching his barking opposition. The boy ran on a bit further, and then he came to a sudden stop. The dogs were just ahead of him. He had been trying to get away from them, but he had actually been following them! Just then, Blaine saw a small opening in the hedges behind him. Without reservation, he jumped through it, knowing this was his best chance to lose the dogs. As soon as he was through, he saw something that made him happy. Tracey! Tracey was there.

Tracey smiled as she motioned for Blaine to come with her. The twins made their way together through a series of short passages and then found themselves in the longest, straightest passageway they had been in so far.

"I think we lost the dogs," Blaine whispered to his sister.

"I think you're right," Tracey replied. "But we also lost our local expert."

Blaine started to nod in agreement as he looked down the passageway. Instead, his face registered a smile as he stated, "Not quite. Look!"

Tracey looked in the same direction and saw what he saw. Off in the distance stood Fiona McRay, waving her arms in the air, motioning for the twins to come her way. The Sassafrases happily heeded their local expert's call and began running at high speed down the long corridor toward her. As they ran, they wondered if McRay had found the wooden exit door that led out of the maze.

In no time, Blaine and Tracey reached the redheaded botanist, who, instead of saying anything, simply pointed toward the hedge wall. However, where she was pointing, there was no hedge wall, just a rough, uneven opening where someone had evidently hacked through. Though this wasn't the wooden door they'd been looking for, it was an exit. All they could see beyond the crude opening was green rolling hills, a line of trees, and, thankfully, no hedges.

Fiona jumped through the opening first, followed closely by the twins, and the three of them ran through the hedge-less field toward the line of trees. "I can still hear the dogs barking," Fiona said as she ran. "So let's get to those trees and out of sight before they find us."

The Sassafrases agreed wholeheartedly with the botanist's plan, and the group traveled together until they made it to the safety of the trees. When they reached the wooded area, they hunched down and looked back toward the hedge maze. Though they could still hear the terriers barking, there was no visible sign of Duke and Earl.

"They must be turned around in the maze," both twins thought. McRay, satisfied that they weren't being followed, crept deeper into the trees, where the twins could hear the sound of moving water. The Sassafrases followed the botanist until she stopped by a beautiful little stream. The clear water wound its way through the trees and over rocks, gurgling gently as it rolled.

"This is Mossy Rock Creek." Fiona introduced Blaine and Tracey to the moving body of water. "It's a beautiful little stream that runs clear from a mountain spring all the way down to yonder loch. It derives its name from the many mossy rocks that it weaves around as it moves down."

Blaine absentmindedly stepped out to one of the rocks to see if he could cross the creek. Just as soon as his foot hit the moss-covered stone, it slipped, and the twelve-year-old boy landed with a splash in the creek.

"Mossy rocks are usually quite wet and slippery." Fiona giggled not so secretly with Tracey.

Blaine, dripping wet, stood up and quickly hopped out of the stream. At first, he looked a bit perturbed, but then he started laughing with the girls.

Fiona stopped and caught her breath before saying, "Mosses

tend to grow in damp places such as a temperate forests that receive a fair amount of rainfall. They also grow in tropical forests and boggy areas. Mosses belong to the Bryophytes group, which are non-flowering plants that grow in clumps low to the ground. The green masses, which are soft and spongy, typically grow less than a few inches high. They do not have true roots. Instead, they have rhizoids, which help to anchor the plant but do not play a role in taking up water. This means that they can grow on stone walls, rocks, trees, and the soil on the forest floor."

Blaine imagined what it would be like if he had rhizoids growing out of the bottoms of his shoes. If so, then he wouldn't have slipped on the rock, but then again he would also be stuck to the rock, unable to move. Maybe he would just leave the rhizoids to the moss.

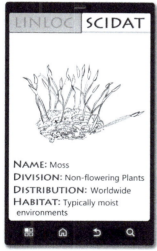

"Moss leaves are small and thin." The sound of Fiona McRay's voice broke off Blaine's daydream. "They are arranged spirally around a slender center stem. The plants do not have any vascular tissue to use in transporting water or nutrients. Instead, a moss absorbs the water and nutrients it needs through the small and thin leaves. Because of this, moss is sensitive to water loss and quickly dries out. However, the moss quickly perks back up when it is able to take up water again.

"Each species of moss has a unique pattern of growth. These plants can grow as variations of dense cushions, mats, or loose clumps that can have a feathery appearance. The life cycle of a moss plant happens in two stages. First, the male and female reproductive cells are released and meet. Then the sporophyte grows, produces spores, and releases them. This is typical of non-flowering plants, which reproduce through the use of spores. Spores are very light because they only contain a few cells, so they are easily dispersed by the wind to a new location."

Blaine and Tracey remembered the discussion of spores from their time with Arrio in the Amazon. They each took pictures of the moss on the rocks. They also made precise mental notes of all that Fiona had just told them, so they could enter their data into their SCIDAT apps later.

They knew that if they got any of the facts wrong, they couldn't move from here to their next location. That is just how the zip lines worked. It was a glitch or something that Uncle Cecil and President Lincoln continued to try and fix, but the twins did not really mind. They actually wanted to learn everything their local experts were telling them.

Throughout all the locations they had been to, they had entered their data correctly, except for one, and that had not been their fault. It was when the Man with No Eyebrows had snuck down into Uncle Cecil's basement and jumbled up the data to sabotage them. Thankfully, it hadn't worked!

"Let's make our way down Mossy Rock Creek toward the peat bog," the red-headed, freckle-faced botanist suggested. "I have a surprise for the two of you."

Peat-filled Pathways

Tracey and her dripping brother followed Fiona down a small trail that ran along the creek. In no time at all, the trail jogged out of the trees, away from the creek, and straight into a swampy-looking area.

"Is this the peat bog?" Tracey asked.

Fiona nodded.

"Is this the surprise?" Blaine asked.

Fiona shook her head. "No. If I'm correct, the surprise lies in that small cottage that you can see on the other side. But before I go revealing big surprises, let me tell you about this magnificent bog!"

The Sassafrases nodded, thinking that sounded good. As Fiona began, they each took a picture of the sweeping landscape they saw before them.

"Peat bogs are made from a blanket of peat moss. Over time, the layer of organic material becomes very thick, with a thin layer of living material on the surface that grows over a thick layer of dead and decaying moss plants. These decaying plants are compressed to form peat, which can be dried and used for fuel, otherwise known as sphagnum moss. Peat grows at a rate of only one millimeter per year, so it takes about a thousand years to form one meter of peat! Is that not daft?"

"Totally daft," Blaine replied.

Tracey rolled her eyes at her brother, knowing he had no idea what the word "daft" meant.

"Twelve different varieties of moss can be found in most peat bogs." McRay continued. "These mosses release an acid that kills bacteria in the soil around them. This provides the nutrients they need and allows for other organic things to be preserved in the soil around them. Would the two of you believe that they have found preserved remains of entire animals in a peat bog? And not only animals but humans too?"

The twins' mouths dropped open. It was amazing to think that such complex things were going on in and around what looked like a plain swampy field.

The botanist went on. "The soil in a peat bog is not as nutrient rich as you find in a marsh or swamp. That is because the main source of water comes from rain, which is low in nutrients. However, you can still grow things in a bog. For instance, rosemary and cranberries grow well in bogs."

Tracey looked over the peat bog to see if she could spot any of the plants Fiona had just mentioned. Blaine looked at his feet, wondering if there were any preserved bodies under the soil he was standing on.

"Peat moss is a type of sphagnum moss that grows in cooler climates that receive large amounts of rainfall," McRay said. "Like in Canada and Finland or right here in Scotland. Here we have two types of bogs. One type is the raised bog, which can be found in the lowlands in old lake basins. The other kind is a blanket bog, which can be found in the highlands where the temperature is low and the rainfall is high."

At first, Fiona had seemed fiery and intimidating, but the more time they spent with her, the more the twins liked her. She was a fascinating wealth of botany information. Blaine and Tracey appreciated hearing about the stunning peat bog before them from their local expert, and they were delighted to be enjoying their botany leg just as much as their zoology and anatomy adventures!

"Now, on to the cottage and, hopefully, the surprise, but we can't just go clambering in," Fiona warned. "We must sneak up to the cottage and then burst through the door all at once."

Blaine and Tracey had no idea why this was the botanist's plan, but they would follow her lead. As the three crept around the bog, McRay gave a little history about the cottage. "This was originally the home and workplace of the Dockerty family's first bottler."

"Bottler?" Blaine asked. "Don't you mean butler?"

"No, laddie, I mean bottler. The bottler was in charge of the Dockerty's distillery. Scottish whisky is typically dried by peat fires, which gives it a distinct taste. So, the bottler lived down here near the peat bog. However, what is most interesting about the cottage is that it can be reached from Dockerty Castle through a secret underground tunnel."

The twins again found their mouths hanging open in astonishment. Their trip to Scotland was awesome! Castles, hedge mazes, peat bogs, and underground tunnels. This was so much cooler than Camp Zip Fire!

Fiona continued. "The tunnel starts in the castle's kitchen and runs all the way down here to this peat bog cottage that usually sits idle and vacant. It is scarcely used, as there hasn't been a bottler in years, but if I were a betting lass, I would bet the cottage is not empty right now. As we were trapped in that hedge maze, I let my mind wander through the happenings of last night from every possible angle, and I think I've figured out how the rubies were stolen—and who stole them. If I'm right, that person is hiding in this cottage at this very moment, and that is the surprise I was speaking of."

Blaine and Tracey had been thinking about the stolen jewels throughout their time in the maze, too, but their minds were blank as to who could have stolen them. Could Fiona be right? Was the real jewel thief in this cottage right now? The three crept quietly, close to the ground, until they reached the cottage. Then they stood up and pressed their bodies against the walls on either side of the front door. Fiona held up three fingers and slowly began dropping them one by one, starting a silent countdown.

Three . . . two . . . one.

The botanist turned and kicked open the old wooden front door of the cottage. She jumped in and pointed her finger in accusation, shouting, "Aha!"

The twins followed her in, overflowing with curiosity to see who the thief was, but they didn't see the thief. Instead, they saw the entire pool of possible thieves. All eight who had chased them through the castle last night stood together around a big rectangular butcher block table that stood in the center of the cottage. The small one-room building had a warm and cozy feel to it, despite the astonished looks on the faces of the eight. Fiona looked a little puzzled at the sight of the eight suspects. She had expected the thief would be alone. She stood tall, still pointing, but who was she pointing at?

"I know who stole the twelve rubies and their crystal stems," McRay said. "So now you can all stop accusing me of being the thief!"

"We are not accusing you." Dunmore the butler interjected. "We know who stole the dozen roses. It was the big caped disappearing man!"

"The big caped . . . what? . . . No!" Fiona said stubbornly. "It was no disappearing man who stole those jewels. It was one of you, and I will prove it, right now!"

"You can't prove anything lass," Dunmore countered. "It wasn't one of us. It was the big caped disappearing man. We saw him with our own eyes, right before he disappeared. In fact, he may still be somewhere in the castle. We got scared, so at Lady Dockerty's command, we took the tunnel from the kitchen down here to the cottage, grabbing flowers and food on the way. I wanted it to feel like home for Lady Dockerty if we happened to be here a long time. But now you come in here—"

The fiery redhead walked up and slammed her hands down on the pastry-filled table, interrupting the butler and showing him who was boss.

"You, Dunmore," Fiona stated, staring at the tall gangly man. "You know Dockerty Castle, better than anyone else. You could have hidden the jewels anywhere. You immediately started blaming everyone after the jewels were stolen. Was it because you were trying to shift the focus off yourself? You, Dunmore, could very well be the thief!"

The butler gasped and took a step back at this accusation.

"Let us review." McRay continued, standing up straight. "Lady Dockerty's brand-new rubies complete with crystal stems were stolen last night during a power outage. The darkness lasted long enough for any one of you to steal the jewels and hide them. Let us also keep in mind that in that darkness there was a specific series of sounds—a grunt, bagpipes, shattering glass, a scream, another grunt, heavy footsteps, a creaking door, and, finally, silence."

Fiona McRay looked at the cast of characters in front of her: the lady, the naval officer, the talent connoisseur, the gardener, the baker, the piper, the maid, and finally back at the butler. Her bright eyes seemed to be looking right into his soul.

"Yes, you, Dunmore, could very well be the thief," she paused, "but you're not! You are a little gruff and bossy, but you are no thief. You are as loyal as they come, and Lady Dockerty is lucky to have you as her butler."

Dunmore exhaled and relaxed, before saying, "Of course I am, madam!"

"Lady Dockerty." Fiona looked at the distinguished madam of the estate. "Obviously, you did not steal your own dozen rubies, but someone standing in this cottage did. I will tell you who that person is."

Lady Dockerty nodded. She wanted to know the truth.

"Angus." Fiona pointed at the gardener. "You looked the guiltiest immediately after the robbery with your shears lying on the table where the rubies once stood. I also happen to know that you have motive. You told me yesterday that you would make Lady Dockerty 'pay' because she was going to add to your workload but not to your salary."

The gardener gulped and glanced nervously at his employer.

"And you did get even, didn't you?" Fiona asked. "But not by stealing the rubies. You got even by cutting a big hole in the outside wall of the Dockerty hedge maze. Am I right?"

Angus nodded his head, confirming that the botanist was correct before looking at his feet in shame.

"Leif." Fiona turned her attention to the piper. "You were knocked unconscious during the robbery, so it seems impossible that you could be the thief, but what better way to pull off a jewel heist than to play the role of the unconscious victim, all the while being fully conscious and having the stolen jewels hidden inside your bagpipes. Is that what caused your bagpipes to sound in the darkness? Did you hide the rubies inside your pipes?"

Leif just stared at Fiona and remained silent. This gave the redhead the opportunity to answer her own question. "No. There are no jewels inside your bagpipes. You don't talk, you smell like lentils, and your name isn't even Scottish. You are strange, but you are no thief. You are a piper, and a piper you will always be."

The kilted man stood a little straighter and gave his bagpipes a little puff as if to agree with Fiona.

"Sir Kentalot." Fiona announced as she turned to the old gray-haired naval officer. "You said last night that your honor is irreproachable. But is it really? I have heard your old war story, the one in which you claim to be a hero. At the center of that story is a chest full of rubies. You seem to love rubies, so what would stop you from stealing rubies from your sister? My very own uncle happens to be one of the men who was with you that cold night when the chest of rubies was found, but the way he tells the story, you were not present when the rubies were found. He says that because of some very overactive insides, you had run off to find the toilet just as the chest was actually unearthed. Is that correct, sir?"

Kentalot's face turned red, revealing the truth without words.

"I see your face is flushing," Fiona noted, "but that isn't the only thing that needs to be flushed, now, is it? The sound of the second grunt last night came from you. You grunted because your stomach demanded that you rush out of the room—not because you stole the rubies, but because you had to quickly find a bathroom. Am I right? Shall we call you . . . Sir Flush-a-lot?"

Knowing he'd been found out, the shamed naval officer confirmed Fiona's theory with a nod.

"And then there were three," McRay stated, looking at Miles, Rona, and Osla. "Let's start with the maid," the botanist-turned-detective stated. "You're good at batting your eyes, and smiling beautiful smiles. Surely, you are completely authentic, or are you? You flatter your employer all the time, but is the lady aware that you have a habit of sneaking into her wardrobe to try on her dresses and earrings?"

Everyone gasped and looked at Osla, who was now covering her face with her feather duster.

"Although this habit is peculiar, it's not a true crime." Fiona continued. "And you're no thief, Osla—at least not of rubies. You didn't steal any jewels, but you did steal a heart, didn't you?"

The maid's face remained hidden. But what caught everyone's attention was how Miles suddenly became strangely anxious at the other end of the table. His face was growing red, and he was fidgeting with his hands.

"Miles, son!" Lady Dockerty exclaimed in a worried tone. "What is wrong with you?"

"He's in love, my lady." McRay filled her in. "The scream that we heard last night came from Osla, but it wasn't a scream of fear. It was a scream of delight. When the lights went out, Miles, who has fallen for the beautiful maid, ran over and planted a big smackeroo right on her kisser. Osla squealed, and then the two of them ran out of the room to confess their undying love for each other. Osla has been flattering Lady Dockerty and trying on her clothes, not because she is a thief but because she is in love with her son and hopes to be the lady's daughter-in-law someday soon!"

Everyone looked at Osla and then at Miles. "It's true," Miles burst out. "I am in love with Osla!"

The talent scout rushed to the other end of the table and embraced the maid. The expression on Lady Dockerty's face was a mystery. Was she happy about her son's romance with a maid or not?

The Sassafras twins' minds raced through the cast of characters involved in the mystery of the stolen dozen roses. Dunmore the butler was loyal. Angus the gardener was vengeful. Leif the piper was strange. Sir Kentalot, the naval officer, was a fraud. Osla, the maid, was in love with Miles, the talent connoisseur, and Lady Dockerty had truly been a victim of robbery. That left just one more person: Rona, the baker. And right now, every eye in the cottage stared directly at her. The plump baker was smiling, laughing nervously, and sweating profusely.

"I didn't steal the dozen rubies, lassie," she told Fiona. "Maybe it was you or one of these two children. Or maybe it was the disappearing man!"

"It wasn't me, the children, or the disappearing man. It was you, Rona, and I can prove it."

Chapter 6: The Gaucho with Ten Names

Argentinian Pampas

"The first grunt," the fiery redhead explained, "came from Leif the piper, as you, Rona, knocked him out of the way to get to Angus. The second noise, the sound of the bagpipes, came from Leif as he fell unconscious on his pipes. You then swiped the lopping shears away from Angus the gardener and gave a mighty swing over the table. It was a direct hit on the glass box that held the jewels, causing the third sound of shattering glass. The fourth and fifth sounds were not caused by you, Rona, but they certainly helped.

When Osla the maid screamed in delight upon receiving the kiss from Miles, the talent connoisseur, and when Sir Kentalot the naval officer grunted upon the gurgling of his insides, it covered up the sound of you snatching up the jeweled flowers. We then heard the sound of your footsteps running to the door, which you pulled open before creeping into the hall. These were the sixth and seventh sounds. We know that three others left the room as well, but they were not with you, Rona. You ran alone, with the stolen jewels, down to the kitchen, where you hid your booty amongst the baking goods. You then returned to the grand sitting room after the lights came back on, arriving at approximately the same time as the lovebirds and the flusher. This is where I, Fiona the botanist, was conveniently framed for the robbery because I was not present. I happened to be hiding in the fireplace with these two wee ones.

"Eventually, the eight of you chased the three of us over the bottom floor of the castle. We eventually escaped and painstakingly made our way through the hedge maze. We traversed Mossy Rock Creek and made our way to the peat bog. I knew that you, Rona, would be in this old cottage. After all, the underground tunnel that connects this place to the castle starts in the castle's kitchen, does it not?"

Fiona's question hung in the air as she paused for a breath before continuing. "Rona, after chasing me and coming up empty handed, you returned to the kitchen and grabbed the jewels along with armfuls of pastries, pies, and such. Then, you raced through the tunnel here to the peat bog cottage. You thought you were going to be alone, but the other seven joined you. I assume they joined you out of fear of this caped disappearing man you say you all saw. But now, Rona the baker, you stand here in the middle of us all—the true thief. You stole the dozen ruby roses!"

The plump baker was still sweating profusely. Her face had also turned bright red with either anger or guilt, or maybe both.

"That's quite a polished theory, lassie," the baker retorted to the botanist. "But you still haven't proven anything!"

"You're right," Fiona replied. "But I'm about to."

McRay began sorting through the many baked goods that were piled high on the big butcher block table.

"Rona, everyone around here knows that you are famous for your version of a baker's dozen. Normally, this refers to a count of thirteen, but for you, it is always eleven because you have a habit of snatching one for yourself. When you saw the dozen roses brought in and presented by Miles to his mother, you found yourself in awe. You wanted those jewels. Your thieving hands began to itch. But with these ruby roses, you didn't want to snatch just one. You wanted them all."

"Prove it, then, lass," Rona stammered, completely flustered.

Fiona finally found what she was looking for as she pulled a pie out from the rest and placed it alone at the center of the table. "What kind of pie is this, Rona?"

"Cherry," Rona answered.

"Cherry, huh." Fiona grabbed a fork from a nearby tin that

was full of clear, cooking utensils. She then plunged the fork through the top of the pie, down into its center. The fork hit something hard. Fiona reached in with her fingers and pulled something red out of the pie.

"This must be the most expensive cherry pie ever made," she quipped.

Everyone could see that what Fiona held in her hand was the color of a cherry, but it was not a cherry. It was one of the stolen rubies. A collective gasp escaped from the small crowd.

"If we dig further through this pie, we will fine eleven more rubies." Fiona continued. "And I'm sure that if we will find the crystal stems hidden amongst the cooking utensils in that tin there."

Another collective gasp went up from the group. Rona fell to the ground and began sobbing. "I did . . . I did it," she sobbed. "I couldn't help myself! I stole the dozen roses."

Later that day, Blaine and Tracey found themselves alone in Dockerty Castle in the exact room they had landed in the evening

before. Only this time, they were sitting on a comfortable couch instead of the rigid stone of the fireplace.

They had successfully sent in their pictures and the required SCIDAT information about flowering plants, hedges, mosses, and peat bogs. They had also opened up the LINLOC application and checked where the next location was. They were going to Argentina—longitude -38° 46' 29.93", latitude -64° 4' 31.80", where they would be studying the pampas, grasses, wildflowers, and ombús. They would have the help of a local expert named Felipe Moreno.

Blaine and Tracey were ready to zip. The Sassafrases' minds were abuzz with excitement about what was coming next. They were eager to see what adventures awaited them in Argentina.

Whoosh! The twins zipped from Europe to South America with light swirling, swishing, and soaring all around them. They landed with a jerk, and their bodies slumped down limp. When strength and sight returned, something that was becoming all too familiar awaited their senses. They had landed in a dark, cramped, box-shaped space—again.

"Oh, not again!" Tracey complained. "I like the tree-top landings, the park-bench landings, and the safari-lodge landings. Not these dark, tight, boxy landings."

Blaine grunted in agreement as he shifted his body, trying to rotate to a more comfortable position. When he did, he landed on something that made a familiar musical sound.

"Trace, I think we've landed inside of . . . a piano," the boy gasped.

Tracey shifted her body as Blaine had done, and touched what felt like metal strings. Yes, the sound that they made did sound like a piano. Tracey reached up with her hands and felt that there was a wooden ceiling to this small black box of theirs. She easily pushed the wooden top up. It opened, spilling light into their landing spot. Blaine and Tracey blinked and looked up at the lazy wooden ceiling fans that were slowly circling overhead.

The Sassafras twins both pulled up and peered slowly out into the light. Indeed, they had landed inside a piano. It was old, wooden, and well used, and was pushed up against a wall of what looked like an old western restaurant or saloon. But surely this was not the real Old West? The invisible zip lines didn't function like a time machine. They could not traveled back in time, but what was this place?

The twins were immediately reminded of the time they had landed in Dubai on a street that looked like an old western town. It had turned out to actually be a fabricated row of buildings built in the parking lot in a mall of the city. The twins wondered if this landing was similar to that one.

This place was quiet, but it wasn't empty. There were two or three men in a back corner playing darts, and there was a big hefty bald-headed man behind a bar polishing glasses. It didn't look like any of the men had noticed the twins' landing. So there was a chance they could quietly crawl out of the piano without being seen.

Blaine started to kick a leg up and over to crawl out of the piano, when suddenly someone burst through the double swinging western-style doors up at the front.

"So sorry I'm late, Emilio!" the man apologized with a native Argentine accent as he waved to the bartender.

The big hefty bartender named Emilio nodded, almost like he had expected the man's tardiness and didn't mind a bit. The man rushed right past the bartender and headed straight for the piano. The twins tried to duck back down inside the big stringed instrument, but it was too late. The man had already spotted them. Blaine and Tracey gulped, assuming they were about to get reprimanded in a major way. Instead, a huge smile found its way to the man's face. He ran up to the piano, and took a seat on the piano bench, still smiling widely at the twins.

"Welcome to Cantina de Pampas!" he announced merrily. "My name is Felipe Moreno, and I would love to play a song for the two of you!"

Blaine and Tracey looked at each other with baffled smiles—this piano player was their next local expert? Evidently, he didn't care a lick that they were sitting inside his piano. That was strange but good. The twins quickly exited their landing spot and remained near the instrument.

Felipe started banging happily on the piano keys. He began singing an upbeat song:

> *Cantina de Pampas, pampas, pampas hey!*
> *Pampas, pampas, pampas hey!*
> *Cantina de Pampas yeah!*

The local expert and piano player continued on with this happy song of few words for several minutes. Then he stopped and just smiled at the twins.

"What are pampas?" Tracey asked the joy-filled musician.

"Pampas are temperate grasslands," Felipe answered in a half-talking, half-singing voice. "The three largest temperate grasslands are found in central North America, Southern Asia, and right here in South America! They are also often called by other names,

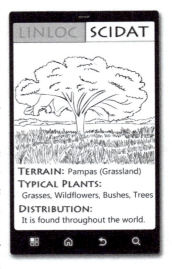

such as the North American grasslands, which are called prairies, the Southern Asian grasslands, which are called steppes, and the South American grasslands, which are called pampas!" At the word "pampas," Felipe excitedly ran his hand from left to right across the keyboard of the piano.

"There are actually two types of grasslands—temperate and tropical. Tropical grasslands are often called 'savannahs,' the largest of which can be found in Africa. Other major tropical grasslands can be found in northern Australia and India. Natural grasslands once covered a fourth of the Earth, but now much of that land is used for farming. Here in South America, the pampas stretch from the Atlantic Ocean all the way to the Andes Mountains, and most of it is located here in the wonderful country of Argentina!" Felipe Moreno paused with a smile and seemed to be sweetly savoring the love he had for his country.

He then continued on with his sing-songy information-giving, "In the pampas, we experience seasons. We have hot summers and cool to cold winters. These pampas contain an abundance of wildflowers and grasses. There are also some low-growing bushes and trees, like the ombú, interspersed. You see, the rainfall is not enough to sustain a large number of tall trees. We get just a little more rain than you would in the desert. Even so, the pampas are teaming with life, such as guanacos, rhea, and burrowing owls."

Moreno paused for another brief moment, and then hit the piano keys again, jumping right back into his song:

Cantina de pampas, pampas, pampas hey!
Pampas, pampas, pampas hey!
Cantina de pampas yeah!

The enthusiastic piano player continued the song as Blaine and Tracey took a seat next to Felipe on the piano bench. They were about to ask him what a guanaco was, when all at once, a mysterious dusty figure stepped slowly through the swinging doors at the front of the cantina. Felipe's piano fell silent, the thud of darts stopped, and Emilio stood frozen with a half-polished glass in his hand. The only sound that could be heard was the soft squeaking of the slow-moving ceiling fans. The mysterious figure stood silently in the doorway. He was an unshaven man, wearing an old leather cowboy hat, a dusty poncho, and worn-out brown boots. He used his eyes to slowly survey the Cantina de Pampas.

"It's a gaucho," Felipe whispered to the twins.

"A guanaco?" Blaine asked.

"No, not a guanaco, a gaucho," Moreno whispered again.

Blaine frowned. He had no idea what a guanaco or a gaucho was.

The eyes of the mysterious man stopped searching and came to a rest on Emilio the bartender. He stepped toward the bar—clump . . . clump . . . clump—his boots sounded on the wooden floor as he walked. He halted at the bar and put both of his rugged calloused hands on the bar-top.

"Hola," the man announced in a deep Spanish voice. "My name is Raul Juan Pablo Eduardo Santiago Mateo De La Casillas . . . the third. I am looking for the man who killed my gray fox."

The glass in Emilio's hand began to tremble. He quickly put it down and stepped back a little.

"Gray fox?" he squeaked out. "I wouldn't know anything about that." The bartender's eyes shifted around nervously. The roughest character around here is Darts Domingo . . . he's there in the back . . . maybe you should go ask him."

The poncho-wearing man stood silent for a moment, still staring at the bartender. He then slowly turned his head, looking toward the back of the cantina, where the small group of men had been playing darts. He lifted his hands off the bar top, turned, and intently strode toward the men.

The three men, whose game of darts had paused, saw him

coming toward them. They quickly turned back to the dartboard, and nervously tried to resume their game.

"Which one of you is Darts Domingo?" the unshaven man demanded softly.

Two of the dart-wielding men immediately pointed to the man standing in the middle of them.

The man looked squarely at Darts Domingo and said, "Hola, my name is Raul Juan Pablo Eduardo Santiago Mateo De La Casillas . . . the third. I am looking for the man who killed my gray fox."

Domingo, who was taller than Raul, shook the nervous look from his face and replaced it with a smug grin.

"I've heard about you," Darts quipped to the man as he twirled a dart around skillfully with his fingers.

"You are the gaucho with ten names." Darts laughed as if he was not impressed. "So, you think I killed your gray fox, huh? Well, maybe I did, and maybe I didn't, but the only way you are going to get that information out of me is if you beat me at a game of darts."

One of Domingo's amigos burst out in laughter. It was clear that the thought of anyone beating Darts Domingo at a game of darts was preposterous to him.

"Your rules or mine?" Raul asked.

"It doesn't matter whose rules we play by, gaucho," Domingo retorted. "Nobody beats me at darts, nobody, regardless of what the rules are."

Raul reached under his poncho and pulled out some kind of coiled-up whip. The whip slowly unfurled until it slapped against the wooden floor. Raul's rugged hand grasped it by the handle.

"A rebenque!" Felipe whispered in awe to the twins.

"A rebenque?" Tracey asked.

"Yes, a rebenque," the piano player replied. "It is the famous leather bullwhip used by the gauchos of Argentina!"

At the sight of the rebenque, Darts Domingo's two amigos took a step back, but he stood his ground, though the nervous expression had partially returned.

"What's your game, gaucho?" he taunted with a slight hint of nervousness.

Raul studied Darts and then glanced at the dartboard. "You throw a dart into the bull's-eye on that dartboard, and then I will knock your dart out of the board using my rebenque. If you miss the bull's-eye, you lose. If I miss your dart, I lose."

"Seriously, gaucho? That's your game?" Darts snorted. "This is going to be easier than I thought. I can hit the bull's-eye all day long."

Wasting no time, Darts Domingo lined his feet up, closed one eye, and took aim. He thrust his arm forward, and—zip—the dart landed with a definitive thud directly in the center of the bull's-eye. He stood up straight, turned, and looked arrogantly at the gaucho.

Domingo's two amigos snickered, while Raul stood still, confident and determined. The twins held their breath expectantly. Darts Domingo's skills were impressive. Would Raul really be able to knock the dart out using a bullwhip? The task seemed next to impossible.

The rugged poncho-wearing man gazed at the dartboard and tightened the grip on the handle of his whip. Then, in one lightning-fast move, he jerked the bullwhip up and out. Smack! The tip of the whip connected with Domingo's dart, knocking it clean off the dartboard. The whole of the Cantina de Pampas sat in stunned and silent awe.

Domingo's face slowly changed from amazement to anger. He grabbed three more darts and zip, zip, zip, threw them all into the dartboard. Three thuds. Three bull's-eyes. Without saying a word, Raul followed with another powerful flick of his rebenque. One smack—and all three of Domingo's bull's-eye darts flew off the board, scattering in different directions.

Now Darts was visibly steaming with anger. He grabbed a handful of darts and took aim at the gaucho with ten names.

Games and Grasses

Ouch! A dart hit him right in the arm.

Ouch again! Another dart hit him in the leg.

He had successfully landed unnoticed in Argentina, in this place called Cantina de Pampas. The Dark Cape suit, complete with the vanish string, was still working. He was invisible, but none of that mattered much to him right now, as he continued to get pelted with metal darts.

Once more, he had been right next to those twins. He had been about to steal their smartphones when something unpredictable had happened. In Peru, it had been a huge native man showing up. In Scotland, he had been attacked by those yappy little dogs. Now here in Argentina, he was getting relentlessly pelted by darts. Would he ever manage to successfully snag the kids' smartphones?

Ouch! Ouch! Ouch! He was hit by three more darts.

That was all he could handle. He re-calibrated his three-ringed carabiner, and once again, he zipped away with his tail between his legs. As he left, he heard the sound of a piano bench crashing to the floor.

Blaine and Tracey sat next to Felipe Moreno behind the overturned piano bench, breathing hard. They had never witnessed anything like what they had just seen. One man throwing darts seemingly as fast as a machine gun can spew out bullets! Another man successfully dodging all those darts, and even knocking some of them out of mid-air with his rebenque.

All was quiet now. The twins followed Felipe's lead as he peeked out from behind the bench that the three had overturned as protection from the darts. There were darts everywhere, in the walls, in the floor, and in the piano bench. Some could even be seen stuck in the blades of the slowly twirling ceiling fans. However, not a single dart could be found on Raul's person. He had dodged them all, and was now standing over the man who had thrown the darts, with the bullwhip in hand. Darts Domingo was lying on the wooden cantina floor in silent defeat.

"So, tell me, Darts, did you kill my gray fox?" Raul asked one more time.

Darts Domingo, whose face was now covered in humiliation, slowly shook his head. "No, Gaucho, I didn't kill your fox, but I have a good idea who did. There is a trapper who lives alone in a shack out on the pampas. His name is Jorge Alfonzo. He is a nasty dirty character who is infamous for collecting animal skins. I bet he is the one who knocked off your fox."

Raul paused and stared into Domingo's eyes. Seeing that he was telling the truth, Raul then slowly nodded.

"Gracias, amigo, for the information." He thanked Darts, reached out a hand, and offered to help Darts up off the floor. Domingo took Raul's hand and stood to his feet as he looked the rugged gaucho over in wonder.

"I have been beaten at darts only one other time in my entire life," Domingo stated. "And that happened just a couple of days ago, but it was a fluke. You see, the man I was playing against cheated. That man was Jorge Alfonzo, and I found out later he was using specially designed magnetic darts. But you, mi amigo, are no cheater and no fluke. You beat me with your rebenque, fair and square."

Raul, in a move that surprised everybody, then held out his whip, offering it to Domingo. Darts looked stunned.

"You're giving me your rebenque?"

Raul nodded. "A token of my appreciation for the information you just gave me about Jorge Alfonzo."

Darts took the whip because he didn't know what else to do. Raul then turned and made his way out of Cantina de Pampas.

Everyone else stood silent and still for a moment until Felipe Moreno broke the silence. "That, mi amigos, was a gaucho!" The piano player exclaimed, pointing toward the swinging doors.

"Gauchos are the pride of Argentina! They are brave and strong and free. They roam the pampas as regal nomads with rich culture and unbreakable integrity. They are a shining symbol against corruption and non-essential material ties. They walk tall, shoot straight, and speak no lies. They are heroes. They are cowboys. They are gauchos!"

Blaine nodded. His gaucho question was answered, but he was still clueless on the subject of guanacos.

Felipe clapped his hands together and smiled. "I am suddenly feeling inspired!"

He set the piano bench back upright, and then sat down, carefully placing his fingers on certain piano keys. He began playing some chords and singing a song that the twins assumed he was making up on the spot.

> *There is a lone gaucho who wanders the pampas.*
> *His face is as stone, his heart heavy as rocks, as rocks, as rocks.*
> *Big pampa sky soars high, but his burden sits low.*
> *As he looks for the man who killed his gray fox, gray fox, gray fox…*

Moreno looked up from the piano with a smile. "That is going to be the chorus," he announced proudly.

THE SASSAFRAS SCIENCE ADVENTURES

"The chorus to what?" Tracey asked.

"To the song that we are going to write about Raul, the gaucho with ten names," Felipe responded. "But we cannot write the song here in the Cantina de Pampas. We must follow Raul out on the pampas! We will accompany him, as guests with music! We will take along several traditional Argentinian instruments and finish the song as we travel!"

At that, Felipe Moreno stood up from the piano and grabbed a hand-crafted wool duffel bag. He headed excitedly toward the front swinging doors and beckoned the twins to follow him.

"Emilio," Felipe said to the bartender as they passed. "Hasta luego, my friend. I'll be back to play the piano for the Cantina de Pampas very soon."

Emilio waved and nodded. The Sassafrases followed Felipe as he raced through the western doors out into the sunlight. The twins had assumed that the Cantina de Pampas was in an Argentinian city—or at least next to a few buildings, but they had been wrong. Once outside, all the twelve-year-olds could see in every direction was miles and miles of silvery-white, windswept grass.

"Ahh, the pampas." Felipe sighed, breathing in the fresh Argentinian air.

"That's a lot of grass," Blaine stated the obvious.

"It sure is," Felipe responded. "What you see before you is silver pampas grass. It can grow eight to twelve inches in height, but how tall it grows is related to the amount of rainfall the pampas receive. So, the more rain, the taller it grows. Grasses, in general, have long slender leaves that help to prevent water loss. Beware of silver pampas grass leaves. They can be razor sharp if you feel along them in the wrong way! They have been known to slice many an unprepared gringo!"

The Sassafras twins used their phones to take pictures of the sweeping temperate grasslands. They also knelt down and got a close-up picture of individual clumps of the silver pampas grass as Moreno continued to divulge information.

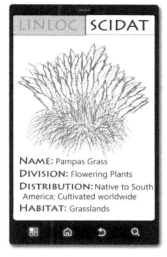

"They grow in clumps called 'tussocks.' When they bloom, it looks like a feathery bush that is white, or even pink. Silver pampas grass blooms are full of seed, which is dispersed by the wind. A typical pampas plant can produce over one million seeds in its lifetime!

"Grasses in the pampas have developed an extensive root system, which reduces erosion and conserves water in the grasslands. You see, fires are often started in the grasslands by people or by lightning. They can quickly spread as the tops of the grasses burn very easily. But because of their roots, the grasses can sprout new shoots within days after a fire. These plants are resistant to damage from fires and grazing because their leaves grow from the base of the plant, unlike the leaves of typical flowering plants that grow on branches." Felipe smiled, inhaled one more huge breath of fresh air, and then hurried off around the corner of the cantina.

The twins followed him, and when they got around the corner, they were surprised at what they saw. There was Moreno standing next to three small horses and a . . . camel? Llama? No, it was a . . .

"Guanaco!" Felipe exclaimed happily.

Blaine and Tracey both smiled. So, this was a guanaco. The horses were equipped with reins, and Moreno was now adding folded-up ponchos to each of their backs to serve as saddles. He hastily piled a few things onto the back of the guanaco, too. So, evidently, they were going to be riding the horses, while the guanaco carried their supplies.

"We will take these wonderful animals out on the pampas in pursuit of the gaucho with ten names," Felipe announced, feeling excited for a new adventure. "I know just where the shack of Jorge Alfonzo is, so I think we can catch up and offer musical assistance to the hero as he pushes forward in his quest!"

Felipe hopped up on one of the horses, as did the twins. Then, the three headed out on the grassland.

After an hour or so, Felipe opened up the big wool bag he had brought, and he pulled out a strange-looking instrument

"This is called an erkencho," he told the twins. "It's an Argentine folk clarinet. It has a single reed with a horn attached to it."

Felipe put the instrument up to his mouth and blew, creating a haunting tune. He started to sing:

Oh, a gaucho's, sad lonesome song
Song of a companion long gone
Who would have the gall or the heart?
Was it that man who played darts?
Had the fox died by his hand?
Did he deserve the reprimand?
An avenger coming with a whip
Would surely sink that Dart's ship
But no, no, no
It wasn't Darts Domingo.

The words of the slow woeful cowboy ballad floated like the wind out across the wide pampas. Blaine and Tracey could now see a figure riding a horse up in the distance. They were sure it was Raul.

Felipe continued with the chorus:

There is a lone gaucho who wanders the pampas.
His face is as stone, his heart heavy as rocks, as rocks, as rocks.
Big pampa sky soars high, but his burden sits low.

As he looks for the man who killed his gray fox, gray fox, gray fox.

The three rode on to the tune of the sad song and the sound of the erkencho. They eventually caught up with Raul just as he reached an old, leaning, wooden shack. As they dismounted, the gaucho hiked up his boot and kicked the door of the shack wide open. Raul stepped slowly into the small one-room shack with his musical posse following closely behind.

It took the twins' eyes a moment to adjust from the brightness of the pampas to the dimness of the shack. When they did, their eyes saw what their noses had already smelled—an unbathed, unshaven trapper, wearing a disheveled straw hat and overalls with no shirt underneath. He was sitting on a bucket with a surprised look on his face.

The gaucho glared at the stinky man, who the twins guessed was Jorge Alfonzo, before introducing himself, "Hola, my name is Raul Juan Pablo Eduardo Santiago Mateo De La Casillas . . . the third. I am looking for the man who killed my gray fox."

The trapper's mouth hung open for a few seconds until he somewhat composed himself. "Well, if it isn't the gaucho with ten names."

Jorge stood up from his bucket and held out his hand offering a handshake. Raul looked at the dirty outstretched hand a moment. Then he looked right back into Alfonzo's eyes and asked, "Did you kill my gray fox?"

"Well, that's no way to treat a host." Jorge pulled his hand back, suddenly becoming very bold. "Why would you ask me such a thing?"

The Sassafras twins thought the answer to that question was obvious as they looked around at the walls of Jorge's shack, which was covered in animal skins.

"I'll tell you what, gaucho," the trapper said as he spat on his own floor. "If you can beat me in a little competition, I'll answer that question about your dead fox."

Raul's glare deepened, and his brow furrowed a bit.

"I caught a handful of vermin last night," Jorge said. "And right now, they are lying dead on a piece of plywood out behind the shack in need of skinning. The competition is this, gaucho. If you can skin ten vermin faster than I can, I'll give you an answer to your question."

The gaucho stood still and gave no response. He looked around at the walls of the trapper's shack, searching for a certain fox skin. The Sassafrases looked, too. They saw the feathers of owls and other birds and the hides of deer, rodents, and even guanacos, but there were no fox skins. Knowing he needed that answer from Jorge, Raul looked at the dirty trapper and nodded.

"Let's get to skinning!"

The sound of the erkencho floated out across the pampas. It was being skillfully played by Felipe Moreno as he persisted in his desire to add musical accompaniment to the quest of the gaucho with ten names. The Sassafras twins didn't know if there was appropriate music for the act they'd just witnessed. Jorge Alfonzo had led them around to the back of his shack where there were dead rodents lying on a piece of plywood. The trapper had separated the rodents—ten on one side and ten on the opposite side of the big board. He had also placed two filleting knives in the center. One for him and one for his challenger—Raul. He had spent a brief moment explaining the simple rules to the competition again, and then he had taken a long moment bragging about how he was the greatest "skinner" who had ever lived.

Like an official, the trapper had counted down, "On your mark, get set . . ."

But before Jorge said "Go," he reached out and grabbed both of the filleting knives, keeping one for himself and throwing the other out into the silver pampas grass, which left Raul without a knife. Only then did Alfonzo shout, "Go!" With a nasty grin, he grabbed his first rodent and started skinning.

"What a cheater," the twins mumbled under their breaths. This wasn't fair! There was no way Raul would be able to win the competition, which meant he wouldn't get the answer he needed about his fox.

Tracey was about to speak up in protest when suddenly Raul reached down to his side and produced a knife from under his poncho. It was two times bigger than the one that had been thrown into the pampas. The blade glimmered in the Argentine sunlight.

"A facon!" Felipe Moreno gasped in wonder.

"A facon?" Blaine repeated.

"Yes, a facon." Felipe nodded. "It's an imposing, sharp-edged knife used by the gauchos of Argentina. These knives literally feed the gauchos. They kill prey, skin the animals, and are even used as an eating utensil." Blaine and Tracey gasped in wonder.

Jorge groaned. The facon had thrown a hitch into his plan to cheat. However, just as his mouth had advertised, he was a fast animal skinner. He was already finished with one rodent and had moved on to his second. How was Raul supposed to catch up?

The gaucho picked up his first rodent, and with a lightning-quick flick of the wrist, he made his first slice. Before the twins knew what was really happening, Raul pulled the skin clean off the small animal. At the sight of this, Jorge's eyes widened, and he quickened his already blistering pace. The race was on, and it was being accompanied by exciting erkencho music. Jorge Alfonzo was fast and his effort was good, but it was no use. The gaucho was quite a bit quicker than the cheating trapper. Within several moments, Raul had all ten of his vermin skinned, and Jorge had only five.

In the wake of his defeat, Alfonzo glared at the gaucho and tightened his grip on his filleting knife. Every muscle in the trapper's body seemed to tighten. Was he going to attack the gaucho?

Chapter 7: Who Killed the Gray Fox?

Wrangling Wildflowers

All at once, Jorge exhaled and relaxed. The angry lines on his forehead disappeared, and his shoulders slumped, as he harmlessly dropped his knife to the ground.

"I tried to cheat you, gaucho," Jorge admitted, looking at Raul. "But you beat me anyway, and for that, I applaud you. I will give you an answer to your question. I did not kill your fox, gaucho, but I think I know who did. There is a certain wrangler by the name of Franco Lorenzo. He roams the pampas like a big bully, pushing anybody he finds out of his way as he wrangles cattle, wild horses, and foxes."

Jorge paused to spit on the ground before continuing. "I have been directly affected by his tricks. Even now, he and his despicable posse occupy my water well. A week or so ago, they rode up with lassoes twirling and guns blazing and told me to leave. I told them the well was mine. That the Alfonzos of the past had dug with their own hands, but that didn't matter a pinch to them. They took it over anyway. Franco did throw a couple of darts my way; he told me it was reimbursement for the well. Whoever heard of that? Who trades their family water well for a few darts? Anyway, I bet Franco Lorenzo is the one who took the life of your fox."

Raul nodded and then held out his hand, offering Jorge the handshake that he had rejected earlier. Then, Raul offered the trapper his facon.

"You're giving me your facon?" the dirty trapper asked in shock.

Raul nodded again. "A token of my appreciation for the information you just gave me about Franco Lorenzo."

Jorge took the knife in awe, shocked that the gaucho would be so generous and kind to him after he had tried to cheat the man. In an attempt to return the favor, Alfonzo packed up some of the skinned rodents and loaded them into Raul's saddlebag as the gaucho mounted his horse to go. Raul slightly pulled down on the brim of his hat in thanks. With a gentle kick to his horse, he headed back off into the pampas, with Felipe, Blaine, and Tracey doing their best to follow him on their tiny steeds.

Moreno put away the erkencho and now pulled out what looked like a folded box.

"Is that an accordion?" Blaine asked.

"Nope, this is a bandoneon. It is very accordion-esque, but it's actually in the concertina family of instruments. Like the accordion, you push and pull at its expandable center, but unlike the accordion, it has buttons to press instead of keys. I love the bandoneon because it floats out a distinct Argentinian sound." Felipe smiled.

Right there on the back of his lumbering horse, Felipe began to play his bandoneon as he added to the ever-growing song of the gaucho.

Out in the silver pampas grass
Lived a trapper, rude and crass.
Was the fox a trophy for this man?
He skinned animals again and again
The sad gaucho wanted to know
But the trapper put on a show
Displaying his cheating way of life
So out came the gaucho's knife
But, no no no
It wasn't Jorge Alfonzo.

The Sassafras twins found themselves humming to Felipe's tune as it was becoming more familiar to them.

There is a lone gaucho who wanders the pampas
His face is as stone, his heart heavy as rocks, as rocks, as rocks
Big pampa sky soars high, but his burden sits low
As he looks for the man who killed his gray fox, gray fox, gray fox.

After wandering the windswept grasslands through the late afternoon, the gaucho and the three riders finally found the water well that Jorge Alfonzo had spoken about. Just like he had said, it was occupied by a bunch of wild wranglers. Even now, they were whooping, hollering, spinning lassos, and shooting pistols up in the air. There seemed to be some sort of exciting event happening. They were concentrated mainly in one area, looking into a small corral of some kind. There were so many wranglers that the twins couldn't see what they were so rambunctious about.

Completely undaunted by the ruckus, Raul rode his horse into the small crowd and dismounted. The sudden arrival of a strange newcomer hushed the wranglers. With a clear and resolute voice, the gaucho made known who he was.

"Hola, my name is Raul Juan Pablo Eduardo Santiago

Mateo De La Casillas . . . the third. I am looking for the man who killed my gray fox."

Whispers immediately began to echo among the crowd. "It's the gaucho with ten names," the Sassafrases heard over and over.

All the whispering wranglers turned their heads and looked at one among them, as if he was the head honcho. The man was taller than the rest. He smiled wryly and stepped forward to face the gaucho.

"Are you Franco Lorenzo?" Raul asked.

"I am," the tall wrangler answered.

"Did you kill my gray fox?" The gaucho's question went straight to the point.

Franco, the wrangler, let out a cocky laugh. "You don't beat around the bush, do you, gaucho? I've been accused of many things, but never of fox killing. I guess it does fit the bill, though, doesn't it? I do wrangle foxes from the pampas, and I have quite a reputation of rough-handling. So, you want to know if I killed your gray fox, do you?"

Franco Lorenzo paused and stroked the brim of his fine cowboy hat. "If you want a definite answer, you're going to have to best me in a challenge, gaucho."

The Sassafras twins rolled their eyes upon hearing this—not another challenge!

Franco turned and walked through the small crowd of wranglers to the crude wooden fence of the corral that Blaine and Tracey had spotted earlier. Raul followed him, as did the twins and Felipe, who began playing his bandoneon again, adding a little drama to the unfolding situation.

"Gaucho." Franco pointed to Raul. "This here is our cattle pen, and as you can see, we currently have several bulls roaming around in it. Now, you are probably thinking that this is going to be some kind of bull-riding competition. But that is where you would be wrong, gaucho. We are not bull riders. What we do best is wrangling, and out of all the wranglers on the pampas, I am the best. There is not an animal that Dios has created that I cannot wrangle in record time. So, here is my challenge to you, gaucho: Wrangle a bull faster than me, and I will tell you who killed your fox."

Raul gave a silent nod of agreement. Then he, along with Franco, hopped over the fence into the pen. One of Lorenzo's wrangler buddies threw a couple of lassos, one to Franco and one to Raul.

"You can use the lasso to get a hold of a bull any which way," Franco explained. "Then, you must get the bull down to the ground. Finally, you have to get three of the animal's legs tied up securely together. The first man to do all of this successfully is the winner."

Another cocky laugh came from Lorenzo's throat. "My fastest time is seven seconds. That'll be hard to beat, gaucho, considering it is the fastest known bull wrangling time ever."

Raul's stone-like face remained solemn as he held up the lasso, looked at it, and then dropped it on the ground. From under his poncho, he pulled out one of the strangest-looking things the Sassafrases had ever seen.

"Bolas," Felipe gasped as he momentarily paused his playing.

"Bolas?" Tracey asked.

"Yes, bolas," Moreno confirmed, "the infamous ancient weapon and hunting tool used by the gauchos of Argentina. Bolas are comprised of three leather-bound stones tied together with three foot-long leather straps. A gaucho swings the bolas to his side or over his head. Then, he lets the stones fly. When they make contact with their intended target, they give it a good smack and wrap around the target tightly. Many a running animal has been captured by one of those."

Franco looked at Raul and snickered. "You want to use those

old leather straps and stones instead of a lasso?"

Raul didn't respond. He remained still.

The wrangler shrugged his shoulders. "Fine, gaucho. That will make winning even easier for me."

Lorenzo motioned to several wranglers on the other side of the pen. They were standing near where most of the big bulls congregated. The two took their head honcho's cue, hopped over the fence, and started hollering and slapping at the bulls. This immediately got the big animals angry, agitated, and riled up. They began to quickly spread out around the cattle pen, stomping and sprinting and snorting in irritation.

"All right, gaucho, the challenge starts when I say 'go.'" Franco winked.

In a matter of seconds, the tall wrangler shouted 'go,' and Raul's third duel of the day was on, accompanied by the music of Felipe Moreno. Franco Lorenzo began swinging his lasso in a perfect circle above his head. Then he took off quickly and spryly, running among the angry bulls. He immediately eyed the one he wanted to rope. On his first toss, the wrangler hooked his lasso over the head and neck of one of the bulls. He then jumped on the bull, putting it in a kind of headlock. He grabbed the big animal's horns and twisted. He seemingly effortlessly pulled the beast to the ground. Now all the wrangler had to do to win the competition was tie three of the bull's legs together. This was all looking way too easy for the cocky wrangler.

Raul, still standing where he had dropped the lasso and pulled out the bolas, was now calmly but vigorously swinging the ancient hunting tool above his head. His dark eyes were focused in on one certain angry bull he was aiming to take down. But even if he hit his target on the first try, how was he going to beat Franco in the challenge? The wrangler was already working on tying up the hooves and legs of his bull.

The gaucho flung the bolas straight and strong. The leather-bound stones made a boomerang type sound as they flew through the air toward the angry but unsuspecting animal. Upon contact with the striding bovine, the spinning bolas tangled and wrapped around all four of the animal's legs. The beast immediately fell to the ground, perfectly tied up. Raul had managed to hook, take down, and tie up his bull in only one move!

Franco, who had finished tying up the legs of his bull a solid two seconds after Raul's bull hit the ground, slowly stood and looked at the impressive work of the gaucho. The tall wrangler knew that he had been bested, and his face showed it.

"You did it, gaucho." Lorenzo conceded. "You beat me in the wrangling challenge. For that, I will give you a definite answer to the question that you're asking. I know exactly who killed your gray fox."

Blaine and Tracey both sucked their breath in, in anticipation. Was Raul really about to get the answer he had been seeking?

"Manuel Hernandez," Lorenzo stated clearly. "It was Manuel Hernandez who killed your fox."

Raul's face remained unchanged, but for the first time, there was some hope in his deep Spanish voice.

"Are you sure?"

"I'm positive." Franco nodded, full of certainty. "Manuel Hernandez is my uncle, the brother of my mother. He is a man with an admirable exterior, but I know what he is really like. He illegally manufactures certain metal products, like magnetic darts, and it is he who thrust me into this life of unlawful wrangling. I faithfully go out and get him cattle, horses, and, yes, even foxes. I do everything he asks me to do without question. For this work, he has promised to give me my own small ranch on a plot of land. He promises me this every year, but then every year, he invites me into his study and says 'Franco, my niño, work for me just one more year, and then I will give you your ranch.' Year after year, I never get the ranch that has been promised to me. Meanwhile he amasses more and more wealth. He lives with his wife, Nicoletta, and a house full of servants. It is a grand sprawling ranch out in the pampas called Hacienda de Hernandez."

Franco paused and looked down at the ground.

"When my parents died, Uncle Hernandez took me in. So, it is hard for me to go against him, but it is time. It is time for this illegal wrangling and magnetic dart manufacturing to stop. It is time for him to fulfill his promises to me. It is time for him to be brought to justice. And who better to do that than you, gaucho?"

"I just need to find the man who killed my gray fox," Raul responded. "If that is your uncle, then he is the one I want."

The gaucho dropped to a knee, close to where his bull was lying on the ground, and untangled the bolas from the beast's feet, letting the big animal up. The bull ran away. Raul stood up and held out his bolas, offering them to Franco Lorenzo.

"You're giving me your bolas?" Franco asked full of shock.

"A token of my appreciation for the information you just gave me about Manuel Hernandez." Raul affirmed, giving away yet another one of his marvelous Argentinian gaucho accessories.

The man who had ten names then simply turned, strode over, and mounted his horse, and rode off into the wide pampas. The Sassafras twins and their local expert hurriedly got on their horses and followed him, hoping to be part of his continuing quest.

As they rode on, the twins were overcome by the magnificent beauty of the Argentinian pampas. The sun was setting in the wide sky in front of them, painting beautiful shades of orange, pink, and purple as if on a heavenly canvas. A gentle chilly breeze blew, whipping dreamily at the twins' noses, making the tops of the pampas grass dance. The soft sound of Felipe's bandoneon floated up and down, adding a sweet melody to the scenic evening.

Eventually, the three brought their horses to a stop as they rode up to a small crackling fire glowing on the darkened pampas. Raul had found a suitable camping spot and had already started a fire. He was sitting there, eating some sort of jerky.

Moreno and the Sassafrases dismounted and silently joined the gaucho fireside. Felipe pulled out his bandoneon and added a third verse to the song of the gaucho.

Out by a certain stolen well
The wranglers built a small corral
Where they roped the bulls none so tame
Could the gaucho beat them at this game?
The head honcho was quick and sly
Had he caused the gray fox to die?
His lasso skillfully did sail
Could the gaucho's bolas prevail?
But no, no, no
It wasn't Franco Lorenzo.

There was nothing like it. Sitting around a crackling campfire on the pampas, listening to a wonderfully crafted gaucho ballad. Blaine and Tracey smiled contentedly in the warm orange glow of the dancing flames. They were now so familiar with Felipe's song that the twins began to sing along when he got to the chorus.

There is a lone gaucho who wanders the pampas
His face is as stone, his heart heavy as rocks, as rocks, as rocks
Big pampas sky soars high, but his burden sits low
As he looks for the man who killed his gray fox, gray fox, gray fox.

The Sassafrases liked the song. They hoped, though, that Raul could find this elusive man he sought.

Moreno shared some of the food he had brought from the Cantina de Pampas with the twins. He also showed them a third use for a poncho. Not only could you wear it and sit on it like a saddle, but you could also lay it on the ground to use as a sort of sleeping bag.

The twins fell asleep right away. They slept soundly accompanied by a steady flow of dreams. Blaine was dreaming that the silver pampas grass was coming alive and forming a protective barrier around him. The grass was protecting him from two barking Scottish terriers named Itotia and Itja.

It was working until the dogs started throwing darts at him. The darts were missing, but they were landing in the campfire, causing the embers to spread. "Oh, no!" Blaine thought in his dream. "You can't spread embers around in the pampas! That will start a—"

Blaine woke up with a start. "Wildfire!" he shouted, now fully awake.

It was already morning. The sun shone bright, and the rest of the group was already awake.

"Not 'wildfire,' 'wildflower,'" Felipe gently corrected.

Tracey looked at her brother peculiarly. "Well, good morning, Blaine. Look at this verbena plant we spotted. It's a wildflower, and Felipe was just about to tell me about it as he makes me a tea from the leaves."

Blaine chuckled, a little embarrassed that he had woken up suddenly, shouting "wildfire." To hide his embarrassment, he reached over, found his phone in his backpack, and took a picture of the plant.

He listened as Felipe Moreno said, "Wildflowers are flowers that grow in the wild. In other words, the seeds were not intentionally planted. They are native plants that have not been cultivated. So, wildflowers can grow just about anywhere and are often seen as weeds in more populated areas. Here in the pampas, there are all different colors of wildflowers during the spring and summer." Felipe paused and smiled even bigger.

THE SASSAFRAS SCIENCE ADVENTURES

"Tracey said she had a bit of a sore throat this morning, so I was going to make her a tea from the leaves of this verbena plant. It is known to soothe sore throats and calm the symptoms of a cold," the excited man explained.

"Argentina is in the Southern Hemisphere so it is winter right now in the pampas." Felipe continued. "That is why it was so chilly last night and why we cannot see the beautiful flowers of this verbena wildflower. In the summer, this field would be covered in a purple carpet of blooms. There would be butterflies and other insects flitting from flower to flower as the wildflowers are pollinated by insects. The tiny flowers have five petals and are typically shades of blue, purple, and white.

"As you can see, the verbena has a stiff woody stem with simple leaves that grow on opposite sides. Not only can we use the leaves for a medicinal tea, but we also use them for an essential oil, although the oil made from the northern lemon verbena is far more popular."

Felipe reached down and grabbed the pot off the fire. He poured a glass for Tracey, handed it to her, and offered, "Try this, Tracey."

Tracey blew into the cup to cool it a bit before taking a sip. She let the tea slide down her throat before responding, "Mmmm . . . it is delicious, and my throat is already starting to feel better."

Tracey set her cup down, took out her phone, and snapped a picture of the verbena plant. She picked up her tea again and finished the rest of the soothing liquid.

"Well, now that is taken care of, let us get moving! It looks like our gaucho is already on his way," Felipe said.

Felipe and the twins folded up their ponchos and put them on the backs of the horses, where the ponchos would serve as saddles once again. All three hopped on their petite animals and followed the gaucho with ten names onward into the pampas.

Ombú Occupation

"There it is," Felipe exclaimed. "Hacienda de Hernandez."

The Sassafrases gazed out into the wide grassland in front of them and spotted the same thing their local expert had spotted—a huge house of shining white out in the distance. They had been riding for a good long while now, and they were glad to have found their destination. They wondered what Raul was feeling. They couldn't tell by looking at his face because, just as the song said, his face was as stone. The twins also wondered if they were about to find the man who had killed Raul's gray fox or if the wild-goose chase would continue.

When they reached the hacienda, Blaine and Tracey saw that it was basically a huge mansion right in the middle of the pampas. The house had a long drive, leading straight up to two huge front doors. The house was shiny white because the whole exterior was made of some kind of polished white stone. The hacienda's grounds were outlined with the strangest-looking trees the twins had ever seen. Tracey made a mental note to ask Felipe about this later.

Raul rode up to the portico and dismounted. As he did, a big, well-dressed man walked out of the huge front doors and smiled a greeting.

"Welcome to Hacienda de Hernandez," the man announced. "My name is Manuel Hernandez. May I have the pleasure of knowing who my visitors are on this fine day on the pampas?"

Raul gave the man his answer immediately. "Hola, my name is Raul Juan Pablo Eduardo Santiago Mateo De La Casillas . . . the third. I am looking for the man who killed my gray fox."

"Oh, yes." Manuel nodded as his smile turned to a concerned frown. "The gaucho with ten names. I have heard about you and your plight. I am so sorry that you lost your fox, amigo. Please, please, all of you come in and have a rest."

Manuel Hernandez turned and motioned for the four to follow him inside. Both Blaine and Tracey had curious looks on their faces—this was Manuel Hernandez? He was not acting even close to the way the twins thought he would be acting according to the way Franco Lorenzo had described him.

Manuel led the group of four through a grand white, polished entryway into a comfortable room with cushioned chairs and shiny tables made of more of the white stone. He invited his guests to sit down. He snapped his fingers, and within seconds, the servants had brought five glasses of sparkling iced tea. Hernandez waited for his guests to take sips of tea. Then, he sat down, joined them, and took a sip of his own.

The big man sat back in his chair, looked at Raul with apparent compassion, and asked, "Poor gaucho, do you feel like you are getting any closer to finding the man who killed your gray fox?"

Raul took another sip of his iced tea, swallowed, and opened his mouth to answer when he was interrupted by the sound of a woman's voice calling out from the entryway.

"Oh, Manuel! Manuel! Are you ready to go to the brunch, dear?"

From the room they were in, the twins could see a woman walking down the entry room stairs. She was wearing an elegant dress, high-heeled shoes, plenty of jewelry, and—what was that hanging around her neck? Was that a . . . gray fox fur?

The twins weren't the only ones who had spotted her neckwear. Raul had, too, and he stood up from his chair, staring at the woman as she entered the room.

"Nicoletta! My dear wife!" Manuel said, standing as well

and laughing nervously. "We aren't supposed to leave for the brunch for another half hour." His laugh turned to a glare. "I told you not to come downstairs until I summoned you."

"Manuel, I am just excited about the brunch," Nicoletta answered her angry husband. "So sorry, dear, who are your guests?" she said as she noticed the four.

Raul answered, "Hola, my name is Raul Juan Pablo Eduardo Santiago Mateo De La Casillas . . . the third. I am looking for the man who killed my gray fox, and I believe I just found him!"

Manuel flushed red with guilt and anger. Nicoletta clutched her fox fur as if she would never hand it over. Raul reached for his rebenque, then his facon, then his bolas, but none of them were there.

Manuel Hernandez snapped again, and once more within seconds the servants appeared, but this time they weren't serving tea. They were holding ropes. The four put up a fight, but it was of no use. Minutes later, they found themselves all tied up, being led outside by Manuel and a large group of wranglers who had appeared at Hernandez's call. The Sassafrases didn't recognize any of them from the day before. Evidently, Hernandez had hundreds of wranglers working for him, and the ones they had just been acquainted with were even rougher than yesterday's.

"Go over and tie them to an ombú while I figure out what to do!" Manuel barked to his wranglers.

The rough and dirty men obeyed. Blaine and Tracey soon found themselves sitting under one of the weird trees they had seen earlier, along with Felipe and Raul. They were all tied securely by their waists and hands to the tree's trunk. Several of the wranglers stayed close to watch them, but the four were left all alone for the most part.

Hernandez had gone back inside his hacienda. Most of the wranglers went back to doing whatever he had assigned them to do.

From their vantage point, the twins saw several pens behind the big house, full of horses and cattle. There was also a big cage that looked to be full of foxes.

The gaucho with ten names sat silently with a solemn face. The twins could only assume that he was thinking about how close he had come to avenging his gray fox. To see his poor animal being worn as ornamentation by some rich lady had to have been heartbreaking. Tracey didn't want Raul to dwell on this, so to help occupy his mind with other thoughts, she asked, "Felipe, can you tell us anything about this tree that we're tied to?"

"Sure I can!" the ever-joyful Moreno answered with a smile on his face. "This is an ombú tree. Actually, it's not a tree at all. It just resembles one. It is really a free-standing herb."

"This isn't a tree?" Blaine asked. "Are you sure?"

"Yep, I'm sure." Moreno replied. "It's an herb. Herbs are plants that are used for flavor, medicine, food, or perfume. Typically, you use the leaves of an herb, but sometimes the bark, berries, and roots also have useful properties as well."

"Well, this sure looks like a tree," Blaine said.

"It does," Tracey agreed. "And it is so tall. I thought temperate grasslands couldn't sustain tall trees or big plants like this because of the low rainfall."

Felipe smiled as he recognized the Sassafras girl reciting some of the information he had given them at Cantinas de Pampas. "The ombú is the only tree-like plant on the pampas because it doesn't need a lot of water to survive," he said.

"It's an evergreen shrub that has an umbrella-like shape, making it popular with animals on especially hot or rainy days because it can provide shelter. The wood of the ombú herb is soft and spongy. So soft, in fact, it can easily be cut with a regular knife. This is because the plant uses the 'trunk' for water storage. It grows quickly by thickening, which is another reason it is not known as a true tree. The ombú herb can grow to be forty to sixty feet in height and forty to fifty feet wide. Its leaves are often used in tea as a laxative. The most interesting thing about the ombú is that it emits a poisonous sap. Typically, the sap can harm you only if it is ingested, but it is poison, nonetheless. So you won't see any cattle grazing on ombú."

NAME: Ombú Tree
DIVISION: Flowering Plants
DISTRIBUTION: South America
HABITAT: Grasslands

"Poisonous sap," thought Blaine. He squirmed and looked around a bit to see if he had come in contact with any. Luckily, this ombú plant seemed to be sap free. Blaine wriggled around a little more and was actually able to free his hands from the ropes. The other three followed his lead, and soon they all had their hands free. Even still, their hopes for escaping remained slim. Not only were their waists still secured to the ombú plant, but there were enough wranglers hanging around that it would be next to impossible to get away.

The twins used their newly freed hands to grab their smartphones and take pictures of the ombú herb. Felipe used his hands to grab yet another instrument out of his bag.

"What instrument is this?" Blaine asked.

"This is a bombo leguero," Moreno replied. "It is a drum that is traditionally made from a hollow tree trunk and covered in cured animal skin. It is one of the oldest instruments in human history and is still used all over Argentina."

Felipe started beating rhythmically on the bombo leguero, and the twins thought he would belt out a fourth verse to the song of the gaucho. Before the musician managed a single word, Manuel Hernandez angrily huffed back over to their location in the shade of the ombú.

"So, gaucho, you think you're really brave and smart, don't you? You rode right up to Hacienda de Hernandez and you found the man who killed your gray fox. But what are you going to do now, gaucho? Don't you know who I am? I am Manuel Hernandez. I successfully wrangle the animals I want to take. I have power. I have money! And I do what I want to do. What are you, a lowly gaucho, going to do to me? Are you going to take me down? Are you going to avenge the death of your fox?" the big man taunted.

Raul stared silently.

Manuel kept going. "I'll tell you exactly what's going to happen, gaucho. Since you love foxes so much, you're going to be treated like one. I'm going to throw you and your three amigos into my fox cage!"

The twins' hearts sank. They didn't want to get thrown into a fox cage. If they could somehow manage to enter their SCIDAT data and enable LINLOC to give them the next coordinates, they could disappear from this place and zip off safely to the next location. The reality was though, even if the Sassafras twins already had their next location, they would not zip away from this place. Sassafrases don't quit, and if they zipped away now, they would be quitting on their new friends.

Besides, they had been in tougher situations than this already this summer, and every time, they had stuck it out. They would stick it out here on the Argentinian pampas, too.

Manuel Hernandez and some of his wranglers untied the four from the ombú herb, jerked them up, and started walking them over to the fox cage. Before they could get there, the ground started rumbling, just like the small tremors of an earthquake.

The Sassafrases looked out at the pampas and saw a line of horses with riders stretched across the entire horizon. The horses were galloping their way. Hernandez started barking orders at his wranglers. He ordered them to take up arms and prepare to defend Hacienda de Hernandez against the approaching cavalry. Hernandez's wranglers began to follow his orders, but as the approaching horses and riders got close enough to identify, the men stopped. It was an army of gauchos.

Everybody knows an army of wranglers has no chance against an army of gauchos. Manuel's wranglers turned to run, every last one of them. Now Manuel stood alone with his four captives as the crowd of gauchos thundered up on his property.

Blaine and Tracey's mouths dropped open at the sight of who was leading this charge of gauchos. At the front of the pack were Darts Domingo, Jorge Alfonzo, Franco Lorenzo, and Emilio the bartender. Darts was cracking a rebenque. Jorge was brandishing a facon. Franco was whirling a bolas, and Emilio was holding up an old piece of paper. The gauchos flanking these four were also wielding their own Argentinian weapons. Manuel Hernandez was so overcome with fear that his knees buckled, and he fell to the ground.

"Hola, my name is Darts Domingo," the darts player said to the man now on the ground. "I lost the first and only game of darts in my life because of a cheater who used magnetized darts. They were darts thrown by a trapper that were given to him by a wrangler. They were given to the wrangler by the man who illegally manufactured them—YOU!"

Now the dirty trapper spoke up. "Hola, my name is Jorge Alfonzo. My family's water well was hijacked by a wrangler—a wrangler who was following orders that were given by—YOU!"

Now it was Manuel's nephew's turn. "Hola, my name is Franco Lorenzo. I have been working year after long year for you, doing things that I don't want to do, because of the promise that you would give me my own ranch. But who turned out to be the liar—YOU!"

Emilio dismounted his horse and walked up to the sniveling Manuel Hernandez, holding up his piece of paper. "Hola," greeted the hefty bald-headed man. "My name is Emilio the bartender. I also happen to be a banker. I'm a fine bacce ball player, too, but my point is this piece of paper that I hold in my hand is a deed—a deed to the land and property that all of us find ourselves on at this very moment. It has come to my attention that several years ago when your sister and brother-in-law died, you forged a fake deed that falsely cited you as the owner of this property. The deed I hold here is the original deed, and it shows that the true owner of this property is Franco Lorenzo. This is not Hacienda de Hernandez. This is Hacienda de Lorenzo!"

Upon hearing this, Manuel Hernandez slumped all the way to the ground, facedown, and started crying.

Later that day, the Sassafras twins found themselves in a mansion full of gauchos. Though these men are not known for being expressive, the mood in Hacienda de Lorenzo was quite festive. Darts Domingo was now considered to have an unblemished darts record. Jorge Alfonzo had regained access to his family's well. Franco Lorenzo had not gotten a small ranch, but instead the grand hacienda that was rightfully his.

The army of gauchos had run the cheating, lying, conniving Manuel Hernandez and his wife Nicoletta out of town. Yes, everyone was happy except for the gaucho with ten names. He had found the man who had killed his gray fox, but it hadn't made him feel any better.

Blaine and Tracey had found the time to enter their SCIDAT data. At any time, they could open up the LINLOC app and see where they were going to zip off to next, but they really wanted to stick around for the fourth and final verse of the song of the gaucho.

At this very moment, Felipe Moreno was gathering everyone together. He had his erkencho, bandoneon, and bombo leguero ready. As a one-man band, he was going to present the song of the gaucho for all those gathered here at Hacienda de Lorenzo. With a smile, Moreno addressed the crowd.

"Good evening, gauchos of the pampas! Tonight, I present a special song to you. It's a song I like to call 'The Song of the Gaucho.' I would like to present this song to Raul, the gaucho with ten names."

Felipe motioned to the chair where Raul had been sitting. But when everyone looked, they saw that the chair was now empty. Raul was not there! The gaucho with ten names had disappeared!

Chapter 8: Pirates in Borneo!

Picture-perfect Palms

They found him out back staring into the cage of foxes. The gaucho with ten names actually had a smile on his stone face. He was pointing into the cage at a certain fox.

Felipe ran up to the cage and looked in. "You have got to be kidding me!" he exclaimed. Moreno then turned around with a smile on his face and announced to the others, "That fox fur that Nicoletta Hernandez was wearing was not from the gaucho's fox! His fox is in here, right inside this cage!"

Then, standing right there next to the fox cage, with a crowd of friends and gauchos, Felipe Moreno finished the song of the gaucho:

> *The gaucho's sad forlorn quest*
> *Took him to the house of Hernandez*
> *A man who seemed so nice and neat*
> *But could he really be a cheat?*
> *He'd wranglers at his beck and call*
> *His wife wore a fox fur as a shawl*
> *Had the gaucho finally found his man?*
> *Was his quest truly at its end?*
> *And yes, yes, yes*
> *It was Manuel Hernandez.*

After singing the fourth verse, Felipe entered into a bridge in the song:

> *But the gaucho's fox is alive, alive, alive!*
> *No he's not dead, not dead, not dead!*

After repeating the bridge several times, Felipe Moreno finished the song of the gaucho with the chorus, altered just a bit, for a perfect finish to the song:

There is a lone gaucho who wanders the pampas
His face is as stone, but his heart's no more
like rocks, like rocks, like rocks
Big pampas sky soars high, and his burdens no more
For he found the man who took his gray fox, gray fox, gray fox!

Upon hearing the closure of the quest, the Sassafras twins stepped away from the group as they cheered for Felipe's ballad and celebrated with the gaucho. They took out their smartphones and opened up the SCIDAT app. They hit "SEND," and the data they had previously entered zipped electronically back to the lab.

The twins then opened up the LINLOC app, which gave them the coordinates of longitude 116° 34' 10.8" and latitude 6° 37' 54.8". Apparently, this would drop them somewhere in Borneo where their local expert would be Trisno Kanang. They would be studying palms, carnivorous plants, molds, and parasitic plants.

Zipping and soaring went the Sassafras twins again through a vortex of light to their next scientific location.

Their light-speed journey concluded, and the twins jerked to a stop. Their bodies landed blind and weak in a new unknown spot. As their sight returned, the white light slowly faded into color, accompanied by the return of strength. Their carabiners had automatically unclipped from the invisible zip lines and the twins now found themselves reclining in extremely relaxed positions.

With their senses fully back and operational, the Sassafrases looked around and saw soft white sand, slow rippling waves, a wispy blue sky, and gently swaying palm trees. They were each seated on their own reclining beach chair facing the ocean. Had they really just landed at their next location, or was this just a dream?

Then, to make the perfect setting even better, a waiter of some sort approached them with a tray full of ice-cold drinks.

"Sir, Madame, would you like a glass of lemonade?" asked the waiter.

Each twin nodded as they reached up and grabbed a sparkling glass of lemony refreshment. The smiles on the twins' faces revealed that they were in a state of shock that was half-disbelief and half-elation. They had never had a landing like this before.

Blaine took a sip of his lemonade and let out a sigh with a happy laugh in the middle of it. He put his hands up behind his head as he leaned all the way back.

"Now this is how it should be," he affirmed. "This is the life."

Tracey smiled and agreed. She took virtually the same posture as her brother before saying, "Yeah, this is way better than landing in a cramped piano or fireplace."

CHAPTER 8: PIRATES IN BORNEO!

As Tracey leaned back in her beach chair, she looked up at the swaying palm trees that dotted the beach around them. The one directly above her seemed to dance to the rhythm of the ocean as it cast a comforting breezy shadow over her brother and her.

"These palm trees are spectacular!" the Sassafras girl announced.

"They sure are," a voice that wasn't Blaine's agreed.

It was the waiter, and he was walking back by their beach chairs. "There are about two thousand eight hundred species of palm plants worldwide, with the largest concentration in tropical and subtropical regions. There are a few of these amazing plants in the desert as well."

The dark-skinned waiter, who looked to be in his teens or early twenties, smiled a seemingly genuine smile and introduced himself. "My name is Trisno Kanang. I am a member of the hotel staff here at Pitchers Beachside Resort. Sorry if I interrupted you two. I just overheard you talking about the palms, and they, along with all the other plant species here on the island, happen to be a passion of mine."

"Nice to meet you, Trisno. I am Tracey Sassafras, and this is my brother Blaine," Tracey offered.

Blaine now jumped in with a smile. "Yeah, Trisno, it is very nice to meet you. And just so you know, we like plants, too, and we would love to hear anything and everything you would like to tell us about them."

Tracey nodded, agreeing with her brother as she asked, "What else can you tell us about these palm trees?"

Trisno Kanang, the twins' new local expert, looked as if he appreciated their friendliness. However, he had an edge of nervousness to him. He set his tray down on a small wooden table and took a seat on a beach chair close to the two.

"Well, first off," he said. "I can tell you that palm trees are not actually true trees."

This information came as a surprise to the twins.

"They are not trees because they don't have a secondary thickening." Kanang continued. "Instead, the seedling grows into a squatty version of the palm until it reaches the final width of the trunk. Then it grows upward, making palms easily recognizable monocots. Some stay short and bush-like, but most are tall, slender, and tree-like plants with a crown of fronds at the top. Fronds are the large, tough, leathery leaves of the palm. Each palm can support only a limited number of leaves, so a new leaf will not unfurl until an older one dies. The roots of the palm do not thicken as they mature. Most of these roots grow underground but in sandy soil, like we have here. The roots will show above ground and spread out to help anchor the plant."

As Trisno spoke, both twins looked over at the nearby palm to see that their local expert was pretty much describing the plant perfectly. They took a picture and then set their phones down on the small wooden table.

Trisno now pointed up at some flowers that could be seen on the palm and started talking about them. "The flowers of a palm are contained on a much larger flower head than most plants. Sometimes, this can have thousands of tiny flowers. Once pollinated, the flowers normally form a one-seeded berry, known as a drupe. Once mature, these seeds fall to the ground and sprout to form another palm tree."

Trisno leaned back on his beach chair, relaxing for a couple of minutes with the twins. Then he hopped up, grabbed his tray, and said, "Nice to meet you two. I'd better get back to work. I'm lucky to have this job, and I don't want to lose it. I really don't deserve . . . it's not . . . you see, what I really am is . . . well, I should just be getting back to work."

The young man turned and walked quickly up the beach and over a sandy ridge. Blaine leaned back again, smiled, and sighed as if he was completely satisfied.

"What are you doing, Blaine?" Tracey scolded him. "We have to go where Trisno goes. How are we going to learn about botany if we aren't around our local expert?"

"Oh, c'mon, Trace—look where we are. It's a tropical paradise! Can't we just spend a couple of hours here, chill-axin'?"

"Chill-axin'? What? No, we are here to learn science, not to chill-ax!"

"OK, OK." Blaine sighed again, but this time it was in resignation. "I guess you're right, but I think we can afford just five more minutes here to finish our lemonades and soak in a little sun and sea air. Then, we will go and track down our local expert."

Tracey nodded in agreement, leaned back, closed her eyes, and allowed herself to relax. "Hey, Blaine," she murmured after a minute of silence.

"Yea, Trace?"

"After Trisno gave us the information about palms and started talking about how he had to go back to work, did you think he started acting weird?"

"Weird?"

"Yeah, kind of like he was hiding something, or like he wanted to tell us something, but he decided not to?"

"I don't know. I didn't really notice anything too weird. Besides, Uncle Cecil wouldn't connect us with anybody too shady, would he?"

Tracey shrugged and nodded, assuming her brother was

right. The twelve-year-olds took their last few sips of lemonade, and then headed up over the sandy ridge they had seen Kanang walk over. When they reached the crest, both twins stopped dead in their tracks as their jaws dropped.

In front of them lay the sprawling, manicured grounds of a luxurious five-star hotel. There were walkways winding through gardens of flowers and plants of almost every color imaginable. There was the biggest swimming pool the twins had ever seen, complete with waterfalls, fountains, slides, diving boards, and multiple levels. There were palms that stood up like kings looking out over their paradise. The hotel itself stood behind all of this grandeur with all the elegance and class one would expect at such a resort.

Blaine and Tracey looked at each other with open mouths. They stood there and stared for a moment on the ridge between the beach and the hotel, forgetting why they were even there. Their eyes were drinking in the scenery instead of looking for Trisno Kanang.

The twins' revere was broken by the sound of an older, obviously wealthy couple who were strolling toward the twins.

"Oh, darling! This place is absolutely marvelous," the twins heard the woman say.

"Yes, my jewel, only the best for you on our fiftieth wedding anniversary," the man responded. "If Pitchers Beachside Resort is fit for movie stars and dignitaries, then it is also a perfect fit for you, my jewel."

The couple walked past the Sassafrases and down toward the beach, holding hands. Tracey cooed, "Awww, Blaine, isn't that sweet?"

Blaine nodded, but he was no longer looking at the celebrating couple. He was looking out into the sea, where a boat was approaching. "It is sweet, Tracey. Look how fast it is."

"Fast . . . what?" Tracey saw that her brother was looking at the incoming boat, not the romantic older couple. She laughed before suggesting, "Maybe it's a pirate ship."

"Yeah, maybe it's like the pirates we saw near the South Georgia Island," Blaine joked with his sister. "What were they called again? The P.R.O. Pirates?"

"Yep, P.R.O. Pirates," Tracey replied. "Pirates from the Piracy Resurgence Organization, led by the less-than-frightening Peach Beard."

"Oh, yeah, Peach Beard!" Blaine chuckled. "He didn't turn out to be very intimidating, did he?"

"I don't know who Peach Beard is," a familiar voice interrupted the twins' conversation. "I can assure you, though, the one who is captaining this incoming boat is intimidating."

The twins turned to see Trisno Kanang standing behind them with concern in his eyes. "Come, Blaine and Tracey Sassafras," he commanded with concern in his voice as well. "We must get out of here right now!"

"Why? Are we in danger?" Tracey asked.

"Yes. Grave danger," Trisno replied. "Because that is, in fact, a pirate ship that is coming in."

"A pirate ship?" asked Blaine. "It doesn't look like a pirate ship."

"Trust me, it is a pirate ship," the hotel worker confirmed. "A ship full of pirates who terrorize the islands of Southeast Asia. They aren't like the pirates you have read about in your fairy tales or seen in your cartoons. They don't mosey along in big wooden ships or dress in ruffles and feathers. No, they look just like ordinary people. They choose their boats for their speed. There are no talking parrots or wooden legs or walking the plank with these guys. These modern-day pirates are precise and efficient. They take hostages, rob, pillage, and plunder, and they do it as fast as lightning. Please, you two," Trisno urged. "We have to get out of here!"

He turned and started running toward the hotel. The Sassafras twins quickly followed him, trusting that his words were true. As the three were running toward the hotel, Trisno shouted out words of warning to all hotel guests in the vicinity.

"Pirates are coming! Find a safe place to hide! Pirates are coming!" he trumpeted over and over again.

The three reached the hotel and rushed in through a big door that was being held open by a white gloved doorman. "Pirates are coming!" Trisno shouted to his coworker as he zipped past him with the twins. The doorman's face wrinkled up with questions, but Trisno wasn't sticking around to answer any of them. He led the Sassafrases through the hotel's large elaborate lobby and into a wide hallway. The three sprinted a ways down and cut right through a set of swinging doors into a smaller hallway.

There were several doors with titles on them. Trisno burst through the one marked "Hotel Manager" without hesitation or a knock. The surprised woman inside nearly jumped on top of her desk at the sudden intrusion. She looked at the three, startled and shocked, but her face eased when she recognized Trisno.

"Trisno, I know I've said that I have an open-door policy as hotel manager, but you just about scared me half to death!" The pretty, middle-aged woman, dressed in a hotel work suit, smiled a kind smile, and then asked, "What can I do for you?"

"Pirates are coming," Kanang said with the same edge of concern in his voice as when he had said it all the times before.

"Pirates?" the woman asked.

"Yes, pirates!" Trisno shouted. "By now, their boat has already landed on the beach, and they are beginning to storm the grounds of the hotel. They will take as many hotel guests hostage as they can, and then they will rob, pillage, and plunder until they've stolen anything and everything of value!"

"Trisno, Trisno, slow down," the hotel manager said to her

emphatic employee calmly. "What makes you so sure about all of this?"

"Mrs. Anita, you have to listen to me! Pirates are coming! And we need to find a safe place to hide, and we need to tell all of our guests and employees to do the same!"

The Sassafras twins could tell that Mrs. Anita, the hotel manager, was a nice woman, and that she was trying to take this threat of a pirate invasion seriously, but that she was having trouble doing so. They were, too. The twins had already run into pirates this summer, as they had been reminiscing about earlier. But those pirates couldn't have plundered a pixie.

Mrs. Anita took a deep breath and looked at her frantic employee. "Trisno, I th—," she began to say, but she was interrupted by the sound of screaming out in the hallway.

The hotel manager's eyes widened a bit, and she quickly stood up to listen at the door.

Carnivorous Captors

He had been very close to swiping the twins' phones back at the beach. They had been reclining on their beach chairs, but then he'd realized that although he was invisible, he was still making visible footprints in the sand. This had sent a sudden jolt of paranoia through him. He didn't want to be found out, so he had just froze near the twins. He wanted to steal the smartphones so badly he could taste it, but he was afraid one or both of the twins would open their eyes just as he made a move. Then they would see him making footprints.

He stood there like an invisible statue for several minutes, and as he did, his paranoia slowly subsided. "Who cares if these twins see my footprints?" he thought to himself. "What will they do to me even if they do? Sure, I haven't been able to stop them this summer, but really, I've had the upper hand the entire time. I've secretly zipped to all . . . well, most of their locations . . . and they still haven't caught me or figured out who I am. Why am I so paranoid? Here I stand, invisible in the Dark Cape suit, and their phones are sitting out in the open on that wooden table just inches away. It's my time. I will steal their phones, and in doing so, I will get sweet revenge on my hated enemy, Cecil Sassafras!"

It had been decided. He would make a move for those smartphones, but he had made the decision too late. Just as he was about to lunge for the phones, the twins had stood up from their beach chairs and had headed up a ridge.

"No, not again!" he had screamed silently to himself. "Why am I always too late?"

He followed them, and by the time he had reached their spot, they had reconnected with their new local expert. There was talk of approaching pirates, and before he knew it, the twins had run away with that Trisno Kanang. They had left him standing invisibly by himself.

Panic had suddenly ensued around him. A big, fast boat had landed on the beach. A scrum of big, nasty, scary-looking men had poured out of the boat and onto the sand. They all had weapons, and with loud voices, they started scaring the living daylights out of everyone on the beach. All the people who had been lounging on the beach or swimming in the surf immediately began running for the safety of the nearby hotel.

He found himself again frozen with paranoia on the sandy ridge that separated the beachfront and the hotel grounds. People were racing and screaming all around him, threatening to knock his invisible frame over. He snapped out of his fearful stance and joined the group running through the grounds of the hotel to the doors of the hotel.

He thought to himself, "Maybe that is the plan of these men—to get everyone into the hotel and then do who knows what."

He slowed to a stop, reached up, and wiped an invisible

bead of sweat from his eyebrow-less forehead. "I can't go into that hotel!" he silently thought.

And so he decided. Yet again he would zip away. Once more, he would have to try and steal the twins' smartphones at their next location.

Trisno Kanang had been right. Pirates were coming. While inspecting the source of the screaming, Mrs. Anita had peeked through the swinging doors from the small hallway out into the bigger hallway. What she had seen sent her scurrying back to her office, visibly shaken. Her look to Trisno confirmed that pirates indeed had come!

The twins were now looking at the hotel manager, hoping that she knew what the protocol was for a pirate attack. Her eyes revealed that she had more fear in her heart than confidence. She sat down at her desk with a silent, frozen stare.

"Mrs. Anita! Mrs. Anita!" Trisno shouted his manager's name, concern still heavy in his voice.

"Please, Trisno," the scared hotel manager begged. "Call me Novi, not Mrs. Anita. Like I've told you before, I'm your boss, but I'm also your friend."

Kanang continued. "Novi, we have to get out of your office! This is one of the first places the pirates will come! We have to find a better place to hide!"

Upon this urging, Novi Anita did not move. So, Trisno took that as his cue to take charge of the situation. The young hotel worker gently grabbed his boss's arm and looked straight at her.

"Novi, follow me. We're going to find a safe place to hide!"

Novi nodded. She stood up from her desk and followed

Trisno. He walked over and slowly opened the office door. Kanang saw that the small hallway was clear. He motioned for everyone to follow him, which they did. Trisno took a right out of the door and went down the hall, away from the swinging doors they had just entered. They could hear screaming and shouting echoing up the hallway.

Trisno Kanang moved quickly and precisely as he led them. It was as if he knew the hallways of the hotel like the back of his hand. He knew exactly where he was and where he was going. The four took several turns through smaller hallways and then eventually came out into a bigger hallway. The hallway was adorned with chandeliers hanging from the high ceiling about every fifteen feet. It had soft carpet with beautiful, intricate designs, but most impressive of all was that the hallway was lined with captivating plants. They were being displayed on ivory pedestals all the way down the hall. The colors of the flowers were hypnotic. The fragrance in the hall was intoxicating.

Blaine stopped in front of one of the pedestals and reached forward slowly, as if hypnotized, to touch the plant. Tracey turned toward the plant in a mesmerized trance. She reached into her pocket and slowly took out her phone. Blaine's hand was centimeters away when Tracey's phone clicked. The "click" sound snapped both children out of their temporary daze.

"What are we doing?" the twins asked each other. They had to keep their senses about them. They had to keep up with Trisno and Mrs. Anita.

The twins scurried away from the plants and ran after Trisno and Novi, who were waiting for them at the bottom of a flight of stairs. There was a small waterfall running down one side of the ivory-stepped staircase. The Sassafrases would have loved to stop and stare at it for a while, but they knew that now was not the time.

The group went up the stairs without stopping. At the top of the stairs, Trisno bolted left after passing several elevator doors. He went left again into a small hallway behind the elevators. There was a door marked "Custodial Closet." Trisno yanked the closet door open and quickly pulled the three inside.

The young man closed the door behind them, and the four stood in the dark until Kanang found the light switch and flipped it on. The fluorescent light bulb slowly came to life as all four, tired of running, found seats where they could in the small closet.

"This closet should hide us well," Trisno puffed. "At least it will for now," he mumbled under his breath.

The local expert looked at the twins with some concern. "Why did you two stop in the hallway by those plants? That was very dangerous. We could have been caught."

The Sassafrases nodded, a little ashamed. "Sorry," Blaine apologized. "Something about those plants made us want to stop, look, and touch them. What kind of plants were they anyway?"

"Pitcher plants," Trisno responded. "They were carnivorous pitcher plants. It is why the hotel is called Pitchers Beachside Resort. We display every one of Borneo's fifty pitcher plant species here in the hotel."

"How can a plant be a carnivore?" Tracey wondered out loud.

"Because these plants can actually *eat* insects. The tips of the leaves form tall narrow cups, called a pitcher, which can collect rainwater. As the pitcher forms and matures, it becomes hollow and full of air. Once mature, the tips of these pitchers have a vibrantly colored hood with nectar-secreting glands, both of which attract insects. The downward-pointing hairs and slick surface inside the pitcher force insects

LINLOC **SCIDAT**

NAME: Pitcher Plant
DIVISION: Flowering Plants
DISTRIBUTION: Southeast Asia
HABITAT: Tropical Forests

down into the water, where they drown."

Trisno paused for effect before continuing. "When the insect falls into the pitcher, it stimulates the plant to release enzymes. These along with bacteria help to digest the prey. Then, the pitcher plant absorbs the nutrient-filled liquid through the walls of the pitcher. It is an example of the simplest form of pitfall traps found in plants."

Blaine looked at his hand. He scrunched up his face as he imagined having digestive plant juices all over it. Though his hand wasn't an insect, he was now glad he hadn't touched one of the pitcher plants in the hallway.

"Carnivorous plants get the nutrients they need by catching insects and small invertebrates in their traps. There are three main types of traps that carnivorous plants use—pitfall traps, snapping traps, and sticky secretions. Pitfall traps pretty much all work like the pitcher plant. They trap the insect or small invertebrate in a pit formed by the plant. Snapping traps, like the Venus fly-trap, wait for the prey to land on them. This action causes the plant to snap shut over the insect or invertebrate, which traps them. Then the plant releases enzymes to digest its catch. The plants that use sticky secretions to trap, like the Sundew, have leaves that have a tacky substance on them, which traps the prey. The trapped insect or small invertebrate stimulates the leaves to curl around and digest it."

"Wow," Tracey marveled. "So plants really can be carnivores—amazing."

Trisno Kanang nodded. "All carnivorous plants use the nitrates and other minerals they absorb from the digested prey to survive."

Novi Anita smiled in admiration at her young employee. Then, directed toward the twins, she said, "Pretty smart, isn't he?"

Blaine and Tracey nodded.

"When I hired him, I wasn't sure if it was going to work out.

He was younger than most employees we have. He was a little rough around the edges, and no one knew where he had come from. But in the end he has turned out to be a good hire."

Trisno's face flushed slightly red as his boss complimented him.

"He is the hardest-working and most diligent worker we have ever had at Pitchers Beachside Resort, and on top of that, his knowledge of the local plants is unmatched. Never before has this hotel had someone who knew so much about Borneo and its plant life." Novi beamed like a proud mother.

Trisno shook his head as if he would not accept his boss's compliments. A sudden frown came over his face as he replied, "If you only . . . if you only knew . . . who I . . . You wou—" Trisno's blubbering sentence was cut off by the sound of approaching footsteps outside the closet door.

Novi and the twins froze in fear. Trisno, on the other hand, sprung into action. He jumped up, grabbed a nearby mop, and began to use it to try and barricade the door, but before he could find a suitable way to accomplish this feat, the closet door swung wide open with great force.

Blaine and Tracey shrank back expecting to see nasty, fierce pirates, but that wasn't who greeted them at all. Instead, they saw the love-struck older couple who had been strolling hand in hand on the beach earlier. The couple's faces displayed fear now, rather than loving glances.

Trisno immediately realized they were hotel guests and pulled the couple safely into the closet. Then, he finished the job he had started, successfully finding a way to use the mop to block the door. Both Novi and Trisno did their best to calm and comfort the elderly couple. Once they had relaxed a bit, the man began his explanation of how they had gotten to the second-floor custodial closet.

"Dear friends, thank you so very much for welcoming us into your closet," he greeted with crisp pronunciation. "We are the Ridgeburns. I am Rover, and this is my pricelessly beautiful wife of fifty lovely years, Zaza."

Zaza smiled and waved a delicate hand in salutation.

"We checked into this wonderful resort just this morning. Afterwards, we decided to take a romantic walk on the beach. Suddenly, we were surrounded by armed men shouting at us. One of these unruly characters made a swipe at the string of pearls around my precious wife's neck. That is when she did a lightning-fast front handspring to elude him. I caught him right in the jaw with a spinning back kick, and we escaped quickly to the back of the hotel."

Trisno, Novi, Blaine, and Tracey had been following the old man up to this point, but now they all sat dumbfounded. How could this elderly couple have possibly pulled off an escape from pirates that included a front handspring and a spinning back kick? Surely, Rover Ridgeburn was embellishing the story just a bit.

The Sassafras twins looked at the elderly Ridgeburns sitting on the floor. Nothing about them that could explain how these two had managed to escape the pirates when no one else had. Mr. Ridgeburn was wearing a sharp-looking white suit and Mrs. Ridgeburn a lovely white dress. If they were celebrating their fiftieth wedding anniversary, that meant they had to be in their seventies, or at least late sixties, didn't it?

Trisno opened his mouth to dig a little deeper into the Ridgeburns' story, when, suddenly, a loud voice cackled over an unseen intercom speaker.

"Good morning, guests and employees of Pitchers Beachside Resort," the voice announced.

Even though the statement was friendly in wording, it came across as creepy due to the deep and arrogant voice that had said it. Upon hearing the voice, the six in the closet froze in shock, but Trisno looked like he had heard a ghost.

"My name is Rama," the deep voice said. "I am now presiding as the new president, CEO, chairman, and king of this resort. I like how that sounds. I am the new king of Pitchers Beachside Resort."

The man named Rama paused for a sinister laugh. He continued as the laugh trailed off. "Semantics aside, my men and I are now in complete control of this hotel. Our ambition today is simply to steal everything of value from this hotel and its people. That may cause some to call us pirates, but we prefer to call ourselves opportunists."

Rama laughed again. "For anyone entertaining the idea of calling for help, let me make you aware that we have set a frequency jammer to block all outgoing and incoming cell phone calls. Additionally, the landlines have been severed. For anyone thinking of planning an escape, be aware that the entire perimeter of the resort is being guarded by my well-armed men. This hotel is extremely remote and hemmed in by the vast ocean on one side and a dense forest on the other side. The privacy that you rich people sought when you booked this hotel has made escape virtually impossible, even if we weren't guarding the perimeter."

Rama paused, as the sound of rustling papers could be heard over the intercom. "According to the hotel's registry, we have rounded everyone up except for two guests, Rover and Zaza Ridgeburn, and the all-important hotel manager, Novi Anita."

He stopped for another moment before lowering his voice to the lowest, most intimidating level yet. "Novi, Rover, Zaza—I know you can hear me right now. I know you are all still somewhere in the hotel. I also know with certainty that we will find you. So hide a little longer if you want, or be wise and hand yourselves over to us right now. Either way, we will find you. We will take you hostage, and we will get what we want."

Rama laughed again and then added in a lighter voice,

"Taking over this resort was even easier than I had thought it was going to be. I would love to pat myself on the back for that, but I must give credit where credit is due. This undertaking would not have been possible without one of our own acting as an 'inside man.' He has been posing as a hotel employee, scoping out all the inner workings of this fine establishment and passing all this information on to us. Join me in thanking him now—Trisno Kanang, we couldn't have done this one without you!"

VOLUME 3: BOTANY   155

CHAPTER 9: ESCAPE INTO THE JUNGLE

Marauding Molds

Blaine and Tracey Sassafras couldn't believe it. They were actually climbing up an elevator shaft! Even harder to believe was that Rover and Zaza Ridgeburn were climbing up the elevator shaft with them. The older couple, who were below the twins, were urging them to go a little faster. Climbing up an elevator shaft was something that had never crossed Tracey's mind, but not so with Blaine. This was something that he had always secretly wanted to do.

But the most difficult thing for the twins to believe though was that their local expert here in Borneo was a pirate. When his name had been announced over the hotel intercom by Rama, the pirate king, Trisno had sworn to everyone that he was not really a pirate. He told them that Rama had to be mistaken, but the way that he had nervously said it, coupled with the cold hard fact that Rama knew his name, led the twins to believe that their local expert was lying.

Pirate or no pirate, Trisno seemed to be desperately trying to help them escape. He was certain that it was only a matter of time before their hiding spot, in the custodial closet, would be found. The young man had pulled open a small half-door in the wall, revealing access to the elevator shaft. Since the hotel was only three stories tall, his idea was to climb up the shaft to the roof. They were already on the second floor, so he didn't think it would be too difficult of a climb for anyone.

Now, here they were, climbing the elevator shaft, with Trisno Kanang in the lead, followed by Novi Anita, the twins, and finally the Ridgeburns. Trisno reached a small door, much like the one they had entered through, pushed on it slightly, and cracked it open. Light seeped into the dark elevator shaft.

Kanang peeked from the darkness out into the light, checking to see if the roof was pirate-free. Evidently it was, because he opened the door wide and climbed out of the shaft without pause. The five swiftly, but cautiously followed out onto the bright sunny rooftop of Pitchers Beachside Resort.

Kanang crept over to the rooftop's edge and peeked over the scant two-foot wall down onto the grounds of the hotel. After scanning and assessing the situation for a few moments, he scurried over to the opposite edge. He did the same spying on that side of the hotel. When he was satisfied, he returned to the five standing in a bunch at the center of the rooftop.

"Okay, here is the plan, my friends," Trisno said with a hint of anxiety on his face. "It is going to be dangerous, but it is our best and only option for escape from this rooftop. There are only four men walking the perimeter. If we time it right, we can hopefully escape out into the jungle without being seen. Unfortunately, what we are going to have to do is jump down three stories into one of the hotel swimming pools."

Tracey's mouth dropped open in surprise. Blaine smiled and

nodded. The Ridgeburns remained expressionless. Novi looked mad and betrayed as she glared at her employee and shook her head.

"How can we trust you, Trisno?" she asked in a strong yet trembling voice. "I thought you were one of my best employees, but more than that, I thought you were my friend."

"I am your friend," Trisno said convincingly.

"No, you're not," Novi responded flatly. "You are a spy, a liar, and a pirate. Did you not do everything the man on the intercom said you did? You used us, you used me, to get all the inside information that you needed."

Novi stared deep into the young man's eyes, looking to see if the truth was hidden there. Trisno looked like he was about to respond right away, but he halted, stared back at Novi for a minute, and then looked down at the ground as though ashamed.

"Okay, Mrs. Anita," he nodded slowly, looking up again. "What Rama said is true, but it is incomplete. I was a pirate. I did sail the seas with Rama and his men stealing, robbing, plundering, and pillaging. I did not have a family of my own. So they were, in a way, like a family to me. I did acquire this job at the hotel for the purpose of spying on it."

Trisno paused. He almost looked like he was tearing up a bit. "But after you hired me and I began working here, I got to know you and the rest of the hotel staff. My heart slowly started to change. The Pitchers staff and employees were a real example to me of what a family should look like. You all appreciated me for me, and not just what I could do for you. You valued my love and knowledge of Borneo's plant life, something that was devalued and mocked by the pirates. My aspirations of robbing Pitchers Beachside Resort *with* the pirates turned into deep longing to protect Pitchers Beachside Resort *from* the pirates."

Trisno paused again to sniffle and wipe his nose, before pleading, "Please believe me, Mrs. Anita."

Novi looked at her employee, searching his eyes. Her face momentarily softened as if her heart was filling with compassion, but then she stiffened. She responded, "You betrayed us, Trisno, and I don't know if I can ever believe you or trust you again. Even if your heart really did change, it doesn't matter much now, does it? Rama and his men control the hotel, and the plan to plunder is in process. You stand here asking us to jump off the roof into the swimming pool. That sounds like a great way to get caught or injured to me. Maybe you want us to get caught. Maybe you are still a true pirate. Maybe you are only pretending to be sorry for what you've done."

Trisno shook his head. "Mrs. Anita, I really am sorry, and I really do want to help you and these other four escape. Not only that, I've already contacted the—"

"Spare me your lies, Trisno." Novi interrupted the young man. "I heard that Rama character call me the 'all-important' employee over the intercom. Why is that? Is it because I'm the only one who knows the code to the hotel's safe? You and Rama are going to try and get that code out of me so you can break into it, aren't you?"

"No, Mrs. Anita! I won't let Rama get to you or the safe!" Trisno assured her.

"I just don't know if I can trust you, Trisno!" Novi Anita replied.

"You can trust him, darling," Zaza Ridgeburn interjected. "He is telling the truth, dear Novi." The elderly lady smiled. "I can see it in his eyes and hear it in his voice. He has changed from who he used to be, and he views you like the mother he never had." Zaza's eyes sparkled as she said this, revealing a wisdom and ability to read people that had come with experience and age.

"Yes, I believe you are correct, my jewel," Rover Ridgeburn added. "Additionally, I believe that his plan to escape by jumping into the swimming pool is a good one."

Rover reached out and grabbed his wife's hand as he gazed at the roof's edge.

"Enough talk. Let us put this contrite young man's plan into action. Are you with me, my jewel?"

"Always with you and never against you, my darling," Zaza crooned.

Then, without hesitating, the Ridgeburns ran, hand-in-hand, to the edge, and jumped, their white garments flapping in the wind. The four rushed to the edge of the rooftop to capture a glimpse of the unfathomable. Rover and Zaza soared gracefully through the air and landed in the pool with the almost nonexistent splashes of Olympic divers. They remained under the water for a few long moments before they emerged unscathed. Rover and Zaza got out of the pool and smiled up at the four on the roof. Their jump had gone unnoticed by the pirates guarding the perimeter.

The Sassafras twins looked at each other with nervous excitement. They needed no more urging to be the next ones to jump. The twins took several steps back from the edge, so that they could get some running momentum. Tracey looked over at her brother and asked, "You want to jump hand-in-hand, too?"

"Sure, Trace," Blaine answered as he grabbed his sister's hand.

Both twins took a deep breath and then launched themselves forward. They jumped up in unison, each putting a foot on top of the short border wall and then jumped out with all their might. Three stories doesn't seem very high when you are traveling by elevator or stairs, but when you are jumping off a building into a swimming pool, the distance seems gigantic.

The Sassafrases floated down through the air, doing all they could to keep their bodies straight up and down. They hit the sparkling blue water of the fancy swimming pool at the same time causing just one splash. The jump and the landing had gone exhilaratingly well. As Blaine and Tracey surfaced, they exited the pool as quickly as they could and joined the Ridgeburns.

Now it was Novi Anita's turn. The twins looked up at her as she peeked over the rooftop's edge and wondered if she would jump. The look on Novi's face was not positive. Nevertheless, the hotel manager backed up and prepared to make the leap.

The next thing they knew, she was coming over the edge. It was a decent jump, though the hotel manager did flail her arms and legs a bit as she went down. However, Novi let out an unintentional scream just before she landed in the water below. The Sassafrases hoped that none of the pirates had heard her. She surfaced quickly and as she exited the pool, Trisno jumped.

Kanang nearly beat his hotel manager out of the water. When he got out, he immediately signaled for the five to follow him. The dripping escapees quickly tiptoed away from the swimming pool and made their way to a row of waist-high bushes. They ducked down behind the bushes and followed the row around to the other side of the hotel. They were heading away from the beach side of the resort, toward the jungle side. The row of bushes eventually ended, and they used single manicured trees for cover, one at a time.

Trisno went first, and as soon as he moved from one tree to the next, Novi made her move. Next was Rover, followed by Zaza, next was Tracey, and finally, Blaine brought up the rear. The five were working their way quickly toward the edge of the resort's grounds. Thankfully, none of them had seen any perimeter-guarding pirates. Blaine was feeling good about their escape thus far.

All that changed when he heard a shout behind him. He turned and saw the exact thing he didn't want to see—a pirate! The man was big, mean, and nasty looking. He was armed with a weapon and continued to shout something Blaine didn't understand. He wasn't looking in Blaine's direction, but why was he shouting? Had they been discovered?

The Sassafras boy wasn't going to wait to find out. He darted from the cover of his tree and ran straight for the jungle at the edge of the hotel's property. As he ran, he pointed a finger to show the others a pirate was coming.

Blaine was fast, and he was getting closer to the thick dense green growth of the jungle, but before he could reach its safety, a dark hand reached out from behind a tree and grabbed him. It was Trisno. He signaled for Blaine to be quiet. Trisno pointed a finger of his own, showing Blaine that another pirate was patrolling the jungle's border. This pirate would have clearly seen Blaine if he entered the jungle.

Blaine gulped as Trisno weighed the small group's options. It appeared that they were stuck between two pirates. In a matter of seconds, Trisno had made up his mind as to what the best plan of action was. He silently signaled for everyone to backtrack and jog to the right where there was a small shed.

As a group, they crept from a tree to the back of the shed. The Sassafrases hoped this shed could provide them with the hiding spot they so desperately needed. The two pirates were getting closer, and now they were both shouting.

Quietly circling around to the front of the small structure, the group let out a gasp of dread. The double doors of the shed were locked with chains and a big lock. Their fear lasted only a moment. In the blink of an eye, Novi pulled out a big ring of keys. She quickly flipped through the ring until she found the key she wanted. She unlocked the padlock, quietly loosened the chains, and opened the double doors.

Novi stepped into the shed first as everyone else filed in behind her. Mrs. Anita then quickly closed the doors and reached through the holes in the doors to grab the chains. She tightened them back up and snapped the padlock shut from the outside. Then, she looked at Trisno and gave him her first smile in a while.

"With a locked padlock on the outside, they won't think anyone is in here," she whispered.

THE SASSAFRAS SCIENCE ADVENTURES

Trisno smiled back at his boss and nodded.

The group remained still and quiet as the sound of the two shouting pirates could be heard just outside the shed. As their gruff voices got closer, Blaine realized he still could not understand what they were saying. They must have been speaking in another language. The pirates were so close that the trembling group heard not only their voices but also their footsteps. Suddenly, the shed's doors were being pulled open from the outside.

Oh, no! The two pirates were trying to open the shed's doors! Would they succeed, or would the re-locked padlock hold? Several long terrifying moments of shouting and door pulling passed. The twins were sure they were caught. Then, to the relief of the six inside, the pirates either gave up or became convinced that no one was inside. The group could hear sounds of the pirates walking away.

It took several minutes of complete silence before the six allowed themselves to exhale and relax a bit. As Blaine and Tracey's eyes adjusted to the dimness, they saw that it was a gardening shed. There were rakes, shovels, lawnmowers, water hoses, and more. It looked like whoever the groundskeeper was also kept this shed as his or her office. A small desk in the back corner was covered in dusty books about landscaping and plants. There was also a wooden bowl filled with . . . moldy fruit. As a matter of fact, now that the Sassafras twins looked around and sniffed, they noticed a LOT of mold.

Blaine leaned over and pointed out, "Trace, look—mold."

"Gross!" Tracey whispered fiercely.

Trisno noticed the twins' reaction to the mold and said, "Mold grows together as a colony. The stuff we can see in this shed is just the reproductive structure of the mold. Underneath, there is a network of connected thread-like structures known as mycelium. This is common to the fungi class. Fungi also lack chlorophyll and are unable to make their own food. Instead, they release enzymes, which can decompose dead organic material and absorb the nutrients they need. Instead of seeds, they release spores, which float through the air. If conditions are right, for instance, a damp or humid shed just like this, the mold will grow and complete its life cycle. You can see this whole process right there in that bowl of decaying fruit."

The twins looked at the wooden fruit bowl again and scrunched up their faces. Some of the plant life they were learning about was a little on the icky side in their opinions.

"Outside in the rainforest," Trisno went on, "the conditions for mold are just right. So, it can grow on the forest floor or on decaying trees. Mold spores are present in every environment, and some are known to produce useful chemicals. One of the most famous is *Penicillium notatum*, which is a very powerful antibiotic that kills bacteria without harming the human body. The mold that it produces grows on decaying fruit. However, some molds are dangerous to humans, and in large quantities, they can cause allergic reactions or respiratory problems."

"So, should we stay in this shed very long?" Tracey asked as she snapped a picture.

"Probably not," Kanang answered. "The mold in here could potentially cause problems for us. Plus, the pirates will eventually come back to open this shed. Our best bet is still to escape into the jungle. But don't worry, the dense forest of Borneo is a haven with a huge variety of plants, trees, and orchids, many of which can be found nowhere else in the world.

"Let's open the shed again, quietly. If it is clear, we can head for the forest," Kanang instructed. "There is a police outpost about twenty kilometers to the south of here. If we can reach it, we will be safe. I have already contac—"

"Wait, wait, wait, Trisno," Novi Anita interrupted her employee. The twins had thought the hotel manager trusted Trisno again, but the tone in her voice said otherwise.

"I don't think it's very wise for us to leave the safety of this shed for the wild jungle. Besides, I have never heard of this police outpost you are talking about." She paused and stared at her employee with the same betrayed look she had shown earlier. "How do I know you aren't just dragging us out into the forest to isolate us and steal from us? Am I right in thinking, 'once a pirate, always a pirate?' Maybe you are the pirate of all pirates. Maybe you want anything that these children or this couple, or I have all to yourself. Out there, you won't have to share with your pirate cohorts. Maybe you think you can get me out into the jungle and pry the password to the hotel's safe out of me."

"Maybe I do and maybe I don't," replied Trisno. "Regardless, Mrs. Anita, you are just going to have to trust me."

Running toward the Giant Rafflesia

Blaine and Tracey looked at their local expert. They could not read him, and they didn't know yet if they sided with him. They hadn't decided if Trisno Kanang was changed or if he was still a tricky pirate. Even so, they felt like they had no other choice but to trust him.

"We have to head for the jungle," the twins stated in unison to Novi.

The hotel manager looked at the Sassafrases, then at the Ridgeburns, and then finally at Trisno. Without saying a word, she took out her key ring, found the right key, and opened the doors to the shed.

Trisno peeked outside, "No sign of the two pirates," he replied. "Let's grab some shovels, which might come in handy, and run for the forest."

All six grabbed a wooden-handled shovel and then headed toward the jungle, using the trees for cover. The group moved quickly. With the cover of jungle green only a few strides away, they began to breathe a little easier. That was, until three pirates stepped out to block their way. One of them was obviously recognized by Kanang.

"Rama!" the pirate's name escaped from Trisno's shocked throat.

Rama, who was even bigger than the other two large pirates, smiled arrogantly. He nodded swiftly at the utterance of his name. "Trisno," he responded as he folded his arms and looked the group over.

"Novi, Rover, Zaza, and I assume a couple of grandchildren. I told you we would find you."

Rama now laughed a big, wicked, open-mouthed laugh, revealing three different kinds of teeth—dirty, golden, and missing.

"As I told you before," the big boss pirate said, "We will now take you hostage and steal from you."

Rama made a gesture toward the two armed pirates to grab the escapees, but before they could, something happened that no one present, save two, expected to happen. Zaza Ridgeburn hunched low and spun her right leg around just above the ground in a three-hundred-sixty-degree kick that swept one of the pirates off his feet. Simultaneously, Rover Ridgeburn hit the other pirate high with a shovel to the head, knocking him down as well.

There was a momentary look of confusion on Rama's face that was quickly replaced with anger as he then lunged at the six himself. Before he was able to touch anyone, Rover spun his shovel as fast as a propeller in front of him. He used the tool to hit the pirate in the stomach, chest, and forehead in rapid succession. Rama, the pirate king, fell down to the ground and joined the pirate chorus of groans.

Trisno, Novi, Blaine, and Tracey were shocked at what they had just seen, but they were not about to stand still. The four of them, plus the astonishing Ridgeburns, burst into the jungle at full speed. None of the six looked back to see if the pirates were following, but it was soon evident that they were being followed. They could all hear the angry shouts close behind.

This sound fueled the Sassafras twins to run with all their might. They both wished that all the world's pirates could be a little more like Peach Beard and a little less like Rama. They would have much rather had that skinny little P.R.O. pirate chasing them. Trisno had been correct back on the beach in saying that Rama was more than a little intimidating.

The group raced on with Trisno leading the way. Blaine found himself at the rear again. He noted the speed and agility of the five in front of him as they quickly moved forward through the jungle. It would be difficult for even the quickest pursuer to catch them, he thought to himself.

Sure enough, after five or ten minutes, Blaine could no longer hear the sounds of the pirates behind them. He hoped this was because they were losing them. Regardless, the six bolted forward through the beautiful forests of Borneo.

Both twins noticed as they ran on that there was much more than just green in this lush jungle. The green was dotted with the beautiful colors of different plants and flowers. Additionally, they passed some of the most breathtaking streams, waterfalls, and natural pools that the Sassafrases had ever seen.

After what must have been well over twenty minutes of running uphill, the group of escapees finally slowed their pace. They kept climbing and jogging, but now they went at a much more manageable speed. The twins considered themselves decent athletes, but both were now tired and breathing hard. Novi Anita and Trisno looked extremely tired as well, but the Ridgeburns looked as brisk and spry as spring chickens in their still spotless white apparel. They all jogged onward at this slower pace for a while longer, until finally Trisno stopped in a somewhat open area amid the dense jungle. The four who needed to catch their breath did as all six stayed quiet, listening for any sign of Rama and his men.

"Rama is big, mean, and intimidating," Trisno finally stated. "But he's by no means the fastest pirate I have ever known. I think we have lost him."

That statement brought some relief to everyone. "We can be certain that though he is somewhat slow, he will not stop pursuing until he finds us." The group tensed up once more. Trisno stood up and jabbed his shovel down into the ground. "I have an idea, though, that I think might stop him."

Kanang shared his plan with his companions. After half an hour of digging and constructing, they had something they all hoped would work to stop and capture their pursuers. It was what Kanang had called a "Pitcher Plant Trap." It was designed to work just like the alluring pitcher plants that the resort was named after. The trap had all the major components of an actual pitcher plant. The six had dug a sizeable ditch that represented the hollow air-filled cup of a pitcher plant. They had diverted some stream water into the ditch, filling it partially and making the walls muddy and slick. Then, they had covered the ditch with leaves and branches and topped it off with Novi's work jacket. They hoped that the pirates would spot the jacket and be lured in, just like insects are lured in by the nectar of the pitcher plant.

Trisno was convinced that if Rama and the other pirates fell into this pitcher plant trap, they would not be able to climb out of it. Everyone hoped he was right. The trap reminded Blaine a little of the trap he had been caught in a few days ago in the Amazon, except their trap here didn't have a snare or spikes. Also, it was not designed to hurt anyone. It was built only to capture the pirates.

The running was over. The trap was complete. Now the six waited to see if the pirates were still following them. They waited to see if their trap would work if indeed the pirates did come.

While they were waiting, a slight breeze began to blow through the jungle. The breeze brought with it one of the worst smells either twin had ever smelled. Blaine and Tracey, who were crouched next to Trisno, covered their mouths and noses. They contorted their faces into looks that clearly communicated what they were thinking.

Trisno noticed and chuckled. "That smell is from the giant rafflesia plant. It is commonly found in the higher elevations of the rainforests in Southeast Asia. It is also known as a 'corpse flower' because it emits a scent akin to rotting flesh."

The Sassafrases both smiled and nodded, now understanding what the horrid smell was. Even so, they kept their hands over their noses and mouths to try and block the smell.

"The smell of the giant rafflesia plant attracts flies," continued Trisno, seemingly unbothered by the putrid smell. "These flies are essential for the plant's pollination and reproduction. Once the flower of the plant is pollinated, it forms berries with minute seeds. These seeds are then able to enter a damaged root or a stem of their host plant and reproduce. The rafflesia is visible only when it flowers, but when it does, the bloom is giant! Thus, the name giant rafflesia. They are actually the largest single flower in the world, weighing up to fifteen pounds and measuring over three feet in diameter."

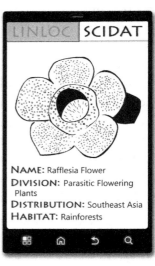

NAME: Rafflesia Flower
DIVISION: Parasitic Flowering Plants
DISTRIBUTION: Southeast Asia
HABITAT: Rainforests

Kanang was still whispering, but he was getting pretty excited. The Sassafrases hoped he wasn't being too loud in case the three pirates were nearby.

"Giant rafflesias are known as parasitic plants." Trisno went on. "Only about one percent of the world's flowering plants are parasitic. Most plants get the water and nutrients they need from the soil and make their own food through photosynthesis. Parasitic plants, on the other hand, get the nutrients and minerals they need from other plants, known as host plants. These plants do not have their own green leaves, which means they cannot make their own food. Many do not have true roots, either."

"There is another kind of plant which is called hemi-parasitic. These plants steal water and minerals from a host plant, but they also have their own green leaves so they can make food. Mistletoe is a good example of this type of plant. Epiphytes, such as orchids, also use the assistance of host plants, but they do so to get closer to the sun, not to steal their food."

Trisno stopped giving information for a moment and started looking around. He then pointed out several other plants within eyeshot.

"There is some mistletoe way up there. Oh! And there is an orchid up high as well. Look! I have spotted a dodder plant. It is a parasitic plant that uses suckers to attach to its host plant. And there—over to the left! There is your smelly culprit. See the giant red and white flower? That is rafflesia!"

The Sassafrases followed Trisno's finger and captured everything that he was pointing out. Their viewing ended with the snap of their smartphone cameras taking pictures of the giant rafflesia. Trisno's joyful excitement ended at the sudden sound of approaching voices.

It was Rama and the two other pirates! They had been successfully tracking the group after all. Blaine, Tracey, Trisno, Rover, Zaza, and Novi all remained completely still, barely daring even to breathe. Would their pitcher plant trap work? It looked like they were about to find out.

Rama, who was leading the pirate trio, spotted Novi's jacket.

He smiled a wicked smile and turned to the two behind him. "See? I told you two numbskulls we were on their trail. Here is that hotel manager's jacket! They can't run from us forever. Nobody escapes from me!"

With his two pirate henchmen at his heels, Rama then stepped forward to grab the jacket. At the precise moment he got his hand on the jacket, the arrogant smile on his face turned into shock because he and his two companions were falling, falling down into Trisno's pitcher plant trap. From their hiding spots, Trisno's group heard screams, a splash, and then more screams. It was surprising that big mean pirates could scream at the octave they were hearing now.

After a couple of minutes of listening to the pirates' screams, grunts, and moans as the men tried to get out of the pit, the six were certain that the trap had worked. With Trisno leading the way, the group came out of hiding, walked up to the edge of the pit, and looked in.

Sure enough, Rama and the other two pirates were hopelessly stuck in the trap. By the looks in their eyes, they were hopelessly stuck in despair as well.

"Don't worry, old boss." Kanang looked down at Rama. "We won't leave you in here forever. We will get some officers from the police outpost to come and fetch you in a bit. I just wanted you to know that I think plundering and pillaging are wrong. It's not a life I want to live anymore. I have found a place where I love to work. It's a place where I can be honest and respected. It's also a place that has people who have treated me like family. And if they'll have me back, that is where I want to stay."

The Sassafrases could tell that Trisno was talking just as much to Novi Anita as he was to Rama. For the first time since right before Trisno's name had been announced on the intercom, the twins saw a complete shift of trust in Novi's eyes. It was clear that she had forgiven Trisno and that she would accept him back as an employee—everything would be okay.

After another hour or so of hiking through the beautifully colored forest of Borneo, the Sassafrases and the Ridgeburns found out that Trisno Kanang had indeed been telling the truth. There really was a police outpost way out in the rainforest, and when they reached it, they learned three more things as well. First, the police officers of Borneo are very efficient in taking pirates into custody. Second, they discovered that before Rama and his men had even landed on the beach, Trisno had contacted the police to tell them about the impending attack on Pitchers Beachside Resort. Even now police officers were in control of the resort, freeing all the hostages and arresting all the pirates. Third, they had discovered that Novi Anita had indeed forgiven Trisno of his trespasses. This forgiveness had been made evident when Trisno tried to turn himself in to be arrested for his wrongs. Novi had spoken up to say that neither she nor Pitchers Resort would be pressing any charges. Instead, the young man was going to get a promotion and a raise. He would be welcomed back with open arms.

With that announcement, Novi and Trisno shook hands. Rover and Zaza gave each other a long kiss and embrace, and the Sassafras twins got ready to zip off to their next location.

Chapter 10: The Secret Siberian Railway

Shrubs and Stowaways

He could see that Cecil Sassafras was as giddy as ever. The twins' SCIDAT information continued to stream in from every location they went to. Even now, pictures and scientific data from their time in Borneo were lighting up the big screen down in Cecil's basement. Cecil and his prairie dog were jumping around joyfully and dancing like fools. It made him sick to his stomach. He watched through his hidden cameras until he got the LINLOC information for the next location, and then he had to look away. He couldn't stand watching his hated enemy celebrate and dance in victory.

Siberia was the next location—longitude 93° 09' 24.5", latitude 55° 56' 06.2". The local expert's name was Pavel Markoff. The subjects of study were dwarf birch shrubs, crocus, lichen, and algae.

If he could just swipe the twins' phones, the Sassafrases would be hopelessly incapable of learning or doing anything. Without phones, they wouldn't be able to log SCIDAT data, take pictures, or progress through the intended locations. He had to get those phones!

To be honest, which he rarely ever was, he was a little ashamed of how timid he had been acting lately. When the twins had first started using the invisible zip lines, he had been much bolder. He had left them marooned among wild animals, blocked them inside a tomb, used a robot hummingbird to spy on them, trapped them inside specially designed boxes, and chased them with rogue robot squirrels. Since he had changed his strategy to the Dark Cape suit, timidity had become too prevalent a theme for him. Sure, the magic suit gave him the ability to become invisible, which should have greatly enhanced his chances to sneak up and steal the twins' phones, but it hadn't quite worked out like that. He had let his paranoia about getting caught control him, but he wasn't going to let that happen any longer.

He had recently remembered that the Dark Cape suit was capable of doing more than simply making him invisible. It could also be used to make objects disappear. He didn't know exactly how it worked, but what if he could get close enough to swing the magic cape over the twins and make *them* disappear? Then his revenge would be complete. No twins equaled no science learning, which equaled a defeated Cecil Sassafras. His dark heart desired that more than anything in the world.

Though their sight and strength had not completely returned yet, they could feel cold wind whipping around their bodies. It let the twins know that they were no longer in a tropical location. Evidently Siberia was a much colder place than Borneo.

"Tracey, I think we've landed on a train," Blaine said as he wobbled to his feet.

"You're right, Blaine." The Sassafras girl shivered as she stood. Her sight and strength had almost fully returned as she replied, "More specifically, it looks like we've landed on the rear platform of a caboose."

The twins looked around at their new surroundings. They were indeed on the back platform of a train, one that was moving with decent speed through a seemingly barren landscape. It was far

from flat, but it was not remarkably mountainous, either.

The twins stuck their heads over the side rails, Blaine to the left and Tracey to the right, to see how long the train that they were on was. It seemed to have only a dozen cars or so with the locomotive engine at the front pulling the train steadily along. The caboose that they were riding was at the very back. The expansive sky was mostly cloudy, and the winds seemed to be quite gusty and cold. However, it was hard to tell how much of that cold and wind was present simply because they were standing outside a moving train.

Blaine was about to make a comment about the cold weather that he hoped would be funny when the back door of the caboose suddenly burst open. A large bearded man dressed in some kind of uniform, which included a gold-trimmed suit, a hat, and a walkie-talkie, immediately started shouting at them.

"Vhat in the name of Belskyduskinhoff are the two of you still doing on the train?" he inquired with a Russian accent. "Everyone vas supposed to disembark in Krasnoyarsk!"

"Sir . . . we didn't—" Blaine began to stammer, but the Russian man interrupted him with a sharp "Nyet!"

The twelve-year-old kept trying to explain. "We didn't know—"

"Nyet!"

"We just want to—"

"Nyet!"

"We landed here—"

"Nyet!"

"I'm sorry, I don't know what 'nyet' means."

"Nyet means 'no.' No, no, no! You two children are not supposed to be on this train!"

The big man grabbed Blaine and Tracey by their shirt collars and pulled them off the back platform, through the door, and into the caboose car. He sat them down and stared at them with an imposing glare.

"The two of you are friends of that stowavay, Sveta, aren't you?"

"Stowavay?" Blaine asked.

"Sveta?" Tracey questioned.

Then both twins shook their heads no. The man looked at the twins a good while longer without saying anything. Then, his glare turned into an apologetic smile accompanied by a sigh.

"OK, children, I can see you are being truthful. Eezveeneete. I am sorry that I vas rough vith you. My name is Yuri Checkoff, and I am the conductor of this train. It vas my job to make sure that all passengers alighted from the train back in the city of Krasnoyarsk. Somehow, I overlooked you two, and for that, I am sorry. I think I may have also missed a scientist vho vas on the train earlier, too. His name is Yuroslav Bogdanovich, and he vasn't even supposed to be on the train in the first place. He is considered to be a very dangerous man. I am fairly sure that I spotted him getting on the train in

Novosibirsk, but I didn't see him get off in Krasnoyarsk.

"Maybe I'm getting too old for this job. Maybe it is time for me to retire; get away from this cold Siberian veather, and just lay on a beach somewhere. I don't know, but I apologize to the two of you. Eezveeneete."

"No sweat," Blaine replied.

"Yes, Mr. Conductor, there is nothing you need to apologize for," Tracey added. "But, if I may ask, where is this train going?"

"I'm sorry, children," Yuri answered. "I am not at liberty to discuss that vith you. It is top secret."

"Top secret?" the twins exclaimed.

"Yes," Yuri replied. "This train is—"

Checkoff's sentence was then suddenly interrupted by the appearance of a man in a Dark Cape. The twins' hearts skipped a beat. In the blink of an eye, two scenarios sped through the Sassafrases' minds.

Number one was that Phil Earp, a magician and friend of the twins, had randomly decided to buy a ticket to Russia and just happened to unknowingly get on the same train as them. Since the train was devoid of passengers, he had decided to use his Dark Cape magician's suit, which gave him the ability to make himself invisible, to surprise them. But why would Phil Earp do that?

Number two was much scarier, but it seemed much more likely to the twins. In this scenario, the Man with No Eyebrows, who had stalked them relentlessly, had stolen the Dark Cape suit again. Now, he was using it to get closer to them, close enough to sabotage their botany studies. The Man with No Eyebrows had proven that he could to travel the globe on the invisible zip lines like them. So, it seemed as if he could be the man in the cape, right?

Seconds after the Dark Cape appeared in the caboose, he raised his cape and lunged at Blaine and Tracey. Scenario number two was looking like the correct one! The twins' minds were racing, but their bodies were frozen like statues. The Man with No Eyebrows flew toward them, but before he could get there, Yuri Checkoff launched himself protectively in front of the twelve-year-old twins.

The wide cape of the magic suit enveloped the conductor. The men scuffled, and the twins heard several thuds and grunts. Then before they could really even tell what was happening, the two men disappeared. Blaine and Tracey stood alone in the caboose with their hearts pounding. They were staring at the spot where Yuri Checkoff and the Man with No Eyebrows had just been. The only thing that remained was Yuri's walkie-talkie lying on the ground.

The twins remained frozen and silent as the train click-clacked on. They would have stayed exactly like that for a good while longer if the walkie-talkie hadn't crackled to life.

"Yuri? Yuri? Are you there, buddy?" a Russian voice asked over the device. "Yuri, look out the vindows on the left side of the train, and you'll see vhat I vas talking about earlier."

Even though neither of the twins were named Yuri, they looked out of the train's left-side windows.

"It's the dwarf birch shrub!" the voice on the walkie-talkie shared excitedly. "They are everyvhere! Can you see them? Their leaves are thick and leathery and are a very dark green on the uppermost surface. That is so they can attract more light for photosynthesis. Their leaves turn bright red and fall off during the short Siberian autumn."

The voice on the walkie-talkie fell silent for a short pause before it crackled back to life with more information about the dwarf birch shrub. "Its flower is called a catkin. It is a slim cylindrical flower cluster that has no petals. It has both male and female flowers that use the vind for pollination. The shrub's root system is much larger and vider than the plant itself because the roots spread out along the thin layer of soil that is just above the permafrost. This type of root system helps to anchor the plant against the vind.

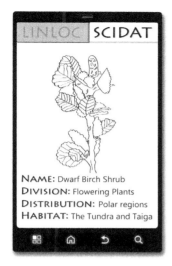

"Did you notice how hairy the shrub looked? Those tiny hairs on the stems help to cut the vind and reduce moisture loss. The coppery-red stems are very tough, vhich helps to resist breakage and splitting. So, the shrub is not crushed or damaged by the veight of vinter snow. The dwarf birch shrub grows in a low clump to avoid this cold vind that you and I are so used to after many years of running this train straight through it. This particular shrub can grow to be about three or four feet tall, and it is found in tundra and taiga regions around the vorld. Just like you and me, old buddy. Yuri? Yuri, are you there?"

The Sassafrases had pinpointed the shrub the voice had been talking about. They had already taken pictures of it, but they were not sure exactly what to do with the walkie-talkie.

Blaine cautiously picked it up off the floor. "Umm…hello?" the boy said hesitantly.

"Yuri?" the other voice replied. "Yuri, you sound strange. Vhat's vrong vith you?"

"This isn't Yuri . . . my name is Blaine, Blaine Sassafras. I am here with my sister, Tracey."

"Blaine Sassafras? Vhat happened to Yuri?"

Blaine paused, pretty sure that what he was about to say would not be believed. "He, ummm . . . disappeared."

"Disappeared?"

"Yes, sir, he got in a fight with a man wearing a magical suit, and they both disappeared."

"A man in a magical suit?"

"Yes, sir."

"It vasn't Bogdanovich, vas it? Yuroslav Bogdanovich? Yuri told me that he thought he had spotted Bogdanovich earlier on the train. That man is crazy! He is a mad scientist and a general danger to society."

There was a slight pause on the walkie-talkie, and then he said, "My name is Pavel Markoff. I am the engineer, and right now I am up at the front in the locomotive, driving the train. If Yuroslav Bogdanovich is indeed on the train, the situation is dire for us all. Vhat I—"

Suddenly, Markoff's voice was cut short, but the walkie-talkie stayed on at his end. The twins listened in fearful curiosity to a series of strange sounds—a clank, a thud, a struggle, groans, a new strange unidentifiable sound, another unidentifiable sound, and then a long eerie booming laugh. Several minutes of silence followed. Then, the twins heard Pavel Markoff's voice again, though it sounded much weaker now.

"Blaine . . . Tracey . . . are you still there?"

"Yes, sir," Blaine responded, with the walkie-talkie still in his hand.

"Yuroslav," Markoff stated in a weak half-whisper. "He is on the train. He just broke into the locomotive and attacked me. He used . . . some strange contraption . . . that he vas holding . . . to tie me up. The good news is . . . he left me still breathing. The bad news is . . . he tied me up so good that I am no longer . . . in control of the train."

The twins could tell by the way that Pavel was speaking that he was hurt.

"Not only . . . am I not in control of the train." The engineer continued. "I also can't reach . . . The brake. So right now this train . . . is unstoppable. And I happen to know that up ahead . . . the tracks are damaged . . . because of a pingo. If ve keep going . . . ve are going to crash."

The Sassafrases didn't know what pingo meant, but they did know what crash meant. The situation didn't sound good. The engineer was tied up, the train was speeding out of control, and there was at least one madman on the loose (Yuroslav Bogdanovich), with the possibility of another (the Man with No Eyebrows).

"Blaine, Tracey, vhere is your current location on the train?" Pavel asked weakly.

"We are in the caboose, sir," Blaine answered.

"It's going to be extremely difficult." Pavel groaned. "But vhat I need you to do is . . . make your vay, from the caboose . . . up to the locomotive. Then I can show the two of you . . .how to stop the train."

Capturing the Crocus

He had missed. He had been aiming for the twins, but that big conductor had gotten in the way. The cape of the suit had wrapped around the big man and had somehow made him disappear.

Aside from the vanish string that when pulled enabled him to disappear and reappear, he still had really no idea how this magic suit worked. He had disappeared along with the bearded man and then reappeared in a different train car all alone with no sign of the big Russian conductor.

He wanted to flee again and simply wait for the twins at their next location, but he would not let himself do that this time. He had resolved to stay at this location. He would work until he had either stolen the twins' smartphones or made the twins disappear. Neither one of those things had happened yet, so he had to make himself stay. He wanted revenge on Cecil Sassafras too much to let anything or anyone prevent him from attaining that ultimate goal.

He stood up from his spot and began moving across the train car. By the time he reached the middle of the car, the door swung open, and in walked the wildest-eyed man he had ever seen. He wore the white lab coat of a scientist, had crazy uncombed red hair, and held in his hand some kind of device that looked a little like a big water gun. The man abruptly stopped and stared at him standing there in the Dark Cape suit, but there was no fear or panic in the man's stare. He only looked curious.

A slow smile broke across the man's face—a wild smile that matched the man's eyes. The smile stayed as the man slowly lifted his gadget and aimed it directly at him. His resolve was gone again. He couldn't let himself get shot with whatever that was. He pulled the vanish string and disappeared.

The Sassafras twins were now out in the cold open air again as they crossed from the caboose to the next car. One slip here could be disastrous. The coupler, the mechanism that holds train cars together, was situated down closer to the tracks. It meant they would have to jump over the coupler gap to get from one car platform to the next.

Both twins managed the jump and made it across without incident. They entered the next car. It looked like a simple boxcar with a few crates here and there.

"Blaine and Tracey." The Sassafrases heard their names on the walkie-talkie.

"Yes, sir?" Blaine responded.

"How is your progress? Any sign of Bogdanovich?" the engineer asked.

"No sign of Bogdanovich, sir, and we are moving toward the front of the train with no problems so far."

"Good, good," Pavel Markoff replied, sounding a little

better. "I've tried to viggle around a little bit . . . and find a vay to get free, but I can't seem to budge an inch. That contraption Bogdanovich used looked like some kind of plastic gun."

"Plastic gun?" Blaine repeated.

"Yes, it vas the strangest thing. After he had me down on the ground, he touched the gun to his shoe, aimed it at me, and pulled the trigger. Now it looks like I am all tied up in . . . shoelaces."

The twins had heard, seen, and experienced some strange things this summer. This contraption that Pavel Markoff had just described was the newest addition to the list.

"Shoelaces?" Blaine asked.

"Da, shoelaces."

Blaine and Tracey worked their way forward out of the boxcar into the next car, which was another boxcar, and as they did, Markoff chattered on the walkie-talkie.

"Yuri and I have been vorking this train and this track together for a long time. If he doesn't reappear, I sure am going to miss him. Strange things have alvays happened on this route, though, maybe because it's top secret. Yuri's not the first one to disappear. Back in eighty-nine, ve had a whole train car disappear right into thin air."

Curiosity formed on the Sassafrases' faces, but they did not voice their curiosity over the walkie-talkie.

"Yes, sir, Yuri and I are pretty good pals." Pavel continued. "His passion vas facial hair, and my passion has alvays been science—botany specifically. It is one of my favorite things about being the engineer of this train. Every day, I get to drive through beautiful Siberia and gaze upon its natural vonders. Right now, ve are rolling through the Siberian taiga, vhich is located just under the tundra. The taiga is the largest biome in the vorld. It has very cold vinters and short, but somevhat varm, summers. There are trees with needles and small shrubs scattered throughout the taiga. Trees here have developed a thick bark to protect themselves against the threat of fire and extreme cold."

The twins now passed to the next car, which was a passenger car.

"Mosses and lichens can also be found in the taiga." Markoff went on. "As can the vonderful crocus plant. Look! Look out of the right side of the train! There are some crocus now. Their flowers are cup-shaped and are typically yellow, white, or shades of purple. Crocus flowers vill open quickly in the spring and early summer. You see them open on sunny days, but vhen the veather is poor, they close up. They vill continue to do this until they are pollinated by insects."

Talking about science seemed to bring life to the train engineer, who was sounding better and better.

"The stem of the crocus is typically a short, thin tube with a large bulb at the base," he told them. "The bulb vill form before the plant goes dormant, and its purpose is to store food for the plant. So that vhen it varms up once again, the flower can grow and reproduce. The visible leaves of the crocus resemble grass that comes from the base of the stem. The plant also has thin papery leaves that cover and protect the bulb. The root system of the crocus plant, vhich anchors it to the ground and connects it to sister plants nearby, is extensive. The crocus is native to scrublands and voodlands from sea level to the alpine tundra and has been cultivated to grow all over the vorld."

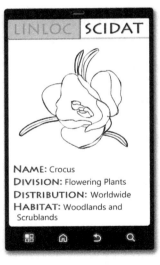

NAME: Crocus
DIVISION: Flowering Plants
DISTRIBUTION: Worldwide
HABITAT: Woodlands and Scrublands

Pavel paused and managed a laugh. "My, my, I can be long-vinded. That is all the botany I vill bore you vith for now, but if the two of you can make it up to the locomotive, I promise to tell you more."

"It's not boring," Blaine assured the man. "We love science and would love to hear anything you would like to share with us."

The twins stopped and made sure they took good pictures of the crocus plant, and then they continued on to the next car—another passenger car. Just as they closed the door after entering, the door at the other end opened, and in walked someone they were immediately certain was Yuroslav Bogdanovich. Their hearts froze in fear. This was the craziest-looking individual either of the twins had ever seen. He had fire engine-red hair exploding from his head, a disheveled white lab coat that hung loosely on his frame, some seriously crazy eyes, and a smile on his face devoid of goodness and full of hubris. His pupils seemed to be pulsating and dilating back and forth uncontrollably.

"Vell, hello, children." The mad scientist took a step forward. He was holding a contraption that looked like a plastic gun that had a nozzle, a handle, and a trigger. However, the body of "the gun" was awkwardly large.

Seeing that the twins staring at the device, Yuroslav took the opportunity to arrogantly inform them what it was. "Impressive, isn't it? I call it the 'Aggrandizer.'"

The twins remained petrified in their places.

"'How does it vork,' you ask?" The crazed man continued on with his one-sided conversation. "You simply place the nozzle of the 'Aggrandizer' on a desired object and then press this button on the side. This causes the desired object to be vacuumed up into the interior capacitator. Then, the brilliantly designed capacitator causes the vacuumed object to multiply and grow exponentially. Next, I pull the trigger and shoot the multiplied object at any given target."

The unmoving twins stared silently.

"You stand in speechless awe at my brilliance, I see,"

Bogdanovich sneered. "You have *heard* how it vorks; now let me *show* you."

The redheaded man stuck the nozzle of his invention up into a high corner of the passenger car.

"I vill vacuum up this spider veb," he said as he pushed the side button, causing the 'Aggrandizer' to make the first strange, unidentifiable sound that the twins had heard earlier on the walkie-talkie.

"Now that the spider veb has grown and multiplied inside the capacitator, it is ready for me to shoot at you, my desired targets." Yuroslav Bogdanovich broke into a laugh that sounded like it was coming from ten men rather than one. Then, he pulled the trigger.

Blaine and Tracey remained helplessly frozen in fear as an unusually large web shot out of the nozzle of the Aggrandizer. The sound coming from the gun now was clearly the second unidentifiable sound they had heard earlier.

Just before they were struck with the spider web, the two were tackled by something else from the side. It was fast and neon green. They landed in a pile behind one of the bench seats with the giant web just missing them.

The Sassafrases sat up quickly and turned to face their tackler. It was a girl. A teenage girl with neon green hair, a neon green electric guitar strapped to her back, a neon green eyebrow ring, neon green pants and shoes, and a black leather jacket with neon green zippers. She was very . . . neon.

"My name is Sveta," the girl announced. "Sveta Corvette."

She pointed in the direction of Bogdanovich, who was still laughing while wildly shooting the Aggrandizer.

"Is that Yuroslav Bogdanovich?" she asked the twins with wide eyes.

Blaine and Tracey nodded.

"Oh, man! I have heard of him!" Sveta exclaimed. "That guy is crazy!"

The twins nodded again. The three stayed behind the seat as Yuroslav kept shooting and shooting and shooting.

He shouted, "I am the greatest scientist this world has ever known! And I vill take control of this train! The train vill be mine! The tracks vill be mine! This top-secret route vill be mine! All of the labs and the research vill be mine! So that one day, all of Siberia vill be mine!"

The mad scientist shot several more spider webs. When he was convinced that he had trapped the children, he left, laughing out the door he had entered.

When the silence had lasted long enough to make them comfortable, the three peeked out from behind the bench seat. They all gasped at the sight of giant spider webs everywhere. The webs virtually filled the entire train car from top to bottom. It was a tangled, impassable mess.

"Despicable," Tracey uttered as Blaine sighed.

"Crazy rad!" Sveta exclaimed.

"Crazy rad?" the twins repeated.

"Yeah, crazy and radical all at the same time!"

"This is no good," Tracey asserted. "We are supposed to be making our way to the front of the train, to rescue the engineer, who is all tied up in shoelaces and unable to pull the brakes to stop the train. It's speeding out of control toward a pingo, whatever that is, where the train will crash! We have to stop this train!"

"Oh, you are right. That is not good," agreed Sveta. "But it is still crazy rad. Crazy rad can be good, and crazy rad can be bad. This is crazy rad bad."

The Sassafrases both had clueless looks on their faces.

"Look, you two," Sveta explained. "My entire life has been

crazy rad. I vas raised by street people in the cold alleys of Moscow, but I alvays kept a positive attitude. Now I am a pseudo-famous punk rocker. I didn't even have a last name, so I just gave myself one: 'Corvette.' The vords to my music are crazy rad good and positively uplifting, just like my attitude. Now I hop trains to travel all over Russia, spreading my crazy rad vibes all across the motherland."

Though she was a little strange, the twins already liked their new friend, Sveta Corvette.

"This is my first time on this train route, though." Corvette looked around the train car. "I have heard of this mysterious train before and the top-secret route that heads up straight north into the tundra. It starts out on the Trans-Siberian railway, but then at Krasnoyarsk, it turns north and travels by itself like a ghost train up through Siberia. I've heard that there are numerous top-secret underground science facilities along this route, vhere scientists do research on all sorts of things."

"Maybe that's why Yuroslav is trying to take over this train," Tracey interjected. "Just think, all those scientists working hard in their labs, and now Bogdanovich wants control of them all. That's what he said, right? But if this madman can grasp control, it would be crazy rad…bad!"

"So we have to stop him!" Blaine joined in with the same amount of growing energy as his sister. "We have to stop this train!"

"But how, Blaine?" Tracey asked. "We can't even move forward through the train car that we're in. There is no way for us to get to the front of the train!"

The twins stared at the hopeless tangle of webs in front of them. While they had eluded the webs of the Aggrandizer, they were still in a tangle of a situation. By the looks of it, there was no way to move ahead with their goal of making it to the locomotive.

"Don't vorry, my friends," Sveta assured them with a sly smile. "There are more vays to move forward on a train than by using the inside of the train cars."

The Russian punk rocker then pointed her finger up.

"The roof?" the Sassafrases asked.

Sveta just smiled and nodded her neon-green head.

Chapter 11: The End of the Line

Look-out for Lichens

Blaine and Tracey had heard Pavel Markoff mention the Siberian wind several times when he had been giving information. They had felt that wind for themselves when they were on the back platform of the caboose. None of that had sufficiently prepared them for what they were experiencing right now.

Facing the Siberian wind from the top of a speeding train was like battling Old Man Winter face-to-face. Even though it was summer here in Siberia, it still felt extremely cold to the Sassafras twins. They could not imagine what it would have been like in winter.

The slick slightly-rounded metal surface of the train top wasn't helping anything, either. Sveta Corvette led the way, followed by Blaine and Tracey as they carefully inched ahead. Everyone was leaning forward, almost to a forty-five-degree angle, afraid that if

they stood straight up, the wind would blow them right off.

The neon-green-haired punk rocker had a smile on her face. This wasn't her first time traversing the top of a train. She considered this activity crazy rad good. The Sassafrases, on the other hand, were rookies. Though they were fans of the bold and daring, they still weren't sure if this was crazy rad good or bad. At this point, they were leaning toward bad. The wind kept blowing, the three kept going, and the unstoppable train click-clacked on.

The twins looked up and saw that the train was approaching a bridge. There seemed to be plenty of clearance for them on top of the train. They wouldn't have to worry about being knocked off, but the bridge was fairly long. It looked like it spanned a deep and wide ravine.

Though the relentless Siberian wind kept beating at Sveta, Blaine, and Tracey, the three were making headway. They got closer and closer to the front end of the passenger car they were on. About the time the train reached the bridge, Sveta made it to the end, and then she carefully climbed down between the train cars. Blaine followed, and then it was Tracey's turn, but the Sassafras girl did not make it. Instead of jumping down in between the train cars, she slipped and began falling off the side of the train car.

The rounded metal roof of the train provided nothing to cling to. Tracey couldn't believe this was really happening. She reached and clawed for a handhold or anything that she could grab onto to stop her from falling. The ravine, which was all Tracey could now see underneath her, was indeed wide and deep. It was going to be a long fall.

Suddenly, her hand caught something—an open window! Tracey had managed to cling to the side of the train by grabbing the edge of the open window with one hand. She quickly got both hands on the window and hung on tightly.

The rattling of the train fought against her, but Tracey was determined not to lose her grip. She caught her breath, prayed a prayer of thanks, and focused on what she needed to do. The window she was holding was the one closest to the front of the car. So, now all she needed to do was reach a hand and a foot over to the platform gap. Then, she could pull herself from off the side of the train and join Blaine and Sveta, who were probably wondering where she was.

Tracey wondered if she could summon the courage to let go with one hand and make the reach for the gap. She was afraid she couldn't. Taking a deep breath, she determined to ignore the fear. She let go with one hand and began reaching. The span was only a couple of feet, but it felt like miles as Tracey inched her trembling hand toward its goal. Her hand and foot made it to the gap at the same time, and letting go of the window seal with her other hand, she pulled herself safely onto the platform. She fell to her side with a gasp and prayed yet another prayer of thanks. She was alive.

Blaine and Sveta had already made the leap to the next train car, and they were was just now entering the other car's door. Blaine looked back to see where Tracey was, and he spotted her lying down. "Tracey! What are you doing? C'mon, make the leap and come with us. We have to keep moving forward."

A look of vexation formed on Tracey's face, letting Blaine know he needed to back off. He quietly turned and stepped into the car. Tracey sighed and rolled her eyes. Evidently, Blaine and Sveta weren't aware that she had almost experienced a fatal fall, but Blaine was right. They had to keep moving forward.

With the accident behind her, the Sassafras girl stood up. She made the leap and entered the next train car, which turned out to be a dining car. Booths with tables lined each side of the car, with the aisle splitting them down the middle. The tables were pre-set with china dishes, silverware, and crisp napkins, as if waiting for the next party of people to come in and dine.

The twins were hungry, and even though they were in a rush, they would have entertained the idea of stopping for a quick bite to eat if it had not been for the individual who was already seated at one of the tables. The three froze as they all saw Yuroslav Bogdanovich. A wicked grin was on his face, and the Aggrandizer was in his hands.

"So, I see that you made it through the vebs. Impressive," he mouthed, sounding neither truly impressed nor concerned.

There was a bowl of something green in front of Bogdanovich on the table. He slowly dipped the nozzle of the Aggrandizer into whatever it was and pushed the vacuum button. The three made a move to turn and back out of the dining car door, but they weren't fast enough. Before they knew it, they were being blasted with a slimy green substance.

Yuroslav laughed with a sinister snicker as he fired his Aggrandizer. The green stuff piled up around the three until it threatened to cover their faces and heads completely. What was this stuff? It was green, slimy, and sticky, but it did have a nice lemon-limey smell.

Blaine, Tracey, and Sveta thrashed at the substance with their arms and legs to try and break free, but it was no use. The stuff had enveloped them, and they were stuck. They could still hear the mad scientist laughing until their heads were covered as well.

The punk rocker and the science-learning twins remained motionless, encased in this see-through green substance. They were almost like prehistoric specimens encased in amber or ice.

Bogdanovich, seemingly satisfied that he had trapped them this time, stopped shooting and exited the dining car.

When Blaine was sure the madman was gone, he immediately began to try and wiggle free. Though this green stuff was thick, he found that he could slowly maneuver around in it. Blaine didn't know if the girls could hear him through this stuff or not, but in case they could, he wanted to let them know it was possible to push through the slime.

He opened up his mouth to shout out the information, but

as he did, some of the green stuff went in. His first instinct was to gag, but he did not because the green stuff tasted very good. It tasted just like green Jell-O!

Blaine smiled as he thought of his two options for escape. He could wiggle his way out, or he could eat his way out!

Slowly, but surely, the Sassafras boy had carved out a sizeable cavity for himself by wriggling and eating. He saw that Tracey and Sveta had done the same. As Blaine dug his way out, the walkie-talkie crackled to life again.

"Are the two of you still alive and kicking?" Pavel Markoff the engineer and local expert asked.

"Yes, sir," Blaine answered as he slurped up another portion of Jell-O. "But we did run into Bogdanovich—two times actually."

"Oh no! You ran into Bogdanovich? I am so sorry. Are you okay?"

"We are okay, but he shot us with spider webs and Jell-O."

"Spider vebs and Jell-O?"

"Yes, sir. So, it seems like he did, in fact, shoot you with shoelaces."

"Yes, I guess he did," Pavel marveled.

"He said that he was trying to take control of the train. Has he come back to where you are yet?" Blaine asked the engineer.

"No, Blaine, he has not, vhich does not make a lot of sense, does it? If he vas trying to take control of the train, one vould think that he vould need to be in the locomotive to do so, but I haven't seen him since he tied me up. You don't think . . ." Pavel's sentence trailed off.

"I don't think what, sir?" Blaine asked.

"You don't think he vants the train to crash do you?" the engineer asked.

With all the train-hopping and Aggrandizer-shooting, Blaine hadn't really thought a lot about Yuroslav Bogdanovich's reasons for doing anything.

"Maybe he knows about the pingo," Markoff continued in his line of thought. "Maybe he really does vant to crash the train."

Regardless of what Bogdanovich's desires were, Blaine knew that they had to stop this train. But he still didn't know what a 'pingo' was, so he asked, "Sir, what exactly is a pingo?"

Always in the mood to talk science, Pavel Markoff had a ready answer. "A pingo is like a frozen hill of sorts," the engineer replied. "They are formed vhen pools of vater under the ground freeze and push up through the surface. Most are fairly small, but they have been known to grow up to three hundred feet in height and half a mile in vidth."

"So that is what a pingo is!" Blaine was satisfied with the knowledge he had just received.

"That's vhat a pingo is," Pavel replied. "And one has formed up ahead directly under the tracks that ve are on right now. It pushed up through the rails, severing the tracks, and ve have not had time to repair it yet. In fact, that is exactly vhat Yuri and I vere on the vay to do. This pingo formed somevhere beyond vhere the taiga turns into the tundra."

From Blaine's current spot, everything still looked green, so he asked, "How is the tundra different from the taiga?"

"The tundra is composed of a thin layer of soil that is less than twenty inches thick," Markoff answered. "It sits over a layer of permanently frozen ground known as 'permafrost.' It is a bitterly cold place for most of the year, but it can come alive during the short summer vith plants and animals. Since there is such a thin layer of soil, the roots of the plants found in the tundra must spread out to avoid the permafrost below. Because of this, you vill not see any trees in the tundra. Mainly, there are just lichens, fungus, mosses, algae, low-growing shrubs, and grasses.

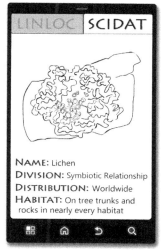

NAME: Lichen
DIVISION: Symbiotic Relationship
DISTRIBUTION: Worldwide
HABITAT: On tree trunks and rocks in nearly every habitat

"Lichens, for instance, grow very slowly in the arctic tundra. Some can grow to be as big as dinner plates and be around one thousand years old." As Pavel continued to give information over the walkie-talkie, Blaine crawled and ate his way out of the Jell-O. It seemed to be getting easier and easier, and the boy had the feeling that he would soon be all the way out. He could see that his sister and their friend Sveta were still moving through the Jell-O as well, so that was good.

The Sassafras boy allowed himself to take a little breather, but it was not wasted time because while relaxing he found a good picture of lichen, using the archive application on his smartphone. He and Tracey hadn't used this app much while studying botany, but man oh man had they used it a lot while studying anatomy. It came in real handy if you were in a situation where you couldn't get a good or actual picture of the subject you were studying, for instance, if you were stuck in Jell-O.

"Lichens are found throughout the vorld," Markoff said. "They can grow on tree trunks and rocks and in some of the harshest environments on the planet. The three main types of lichens are fruticose, foliose, and crustose. Fruticose lichens are shrub-like, foliose lichens are leaf-like, and crustose lichens are flat and crusty. If ve can manage to successfully stop this train and get off, I promise I vill show you some of these amazing plants."

"I would like that," Blaine responded.

"Lichens are really the result of a symbiotic, or mutually beneficial, partnership between a fungus and an algae plant or a cyanobacterium. So they are technically not plants, but rather a living partnership." Pavel continued, the love of botany giving him energy. "The fungus protects the algae or bacteria that live below, vhile the algae or bacteria provide the fungus with the sugars it needs to grow. Lichens reproduce through the use of diaspores, vhich are released in the air. A diaspora is just a spore vith some additional tissue. The lichen diaspores contain spores from the fungus as vell as a few cells from the algae or bacteria."

Splat! All at once, Blaine came tumbling out of the Jell-O at the front end of the dining car. He was slimy, green, and superfull, but he was free. Plus, he knew all about lichens, which was information he would share with his sister later.

Slup! Slurp! Tracey and Sveta oozed out of their sticky green prison, too.

"Okay, Mr. Markoff, sir, we are on the move again and headed in your direction, just as quickly as we can," Blaine announced over the walkie-talkie.

"Thank you, Blaine, thank you," the engineer responded. "I keep viggling around, but it's not vorking. I just can't break free from these shoelaces. So yes, come as quickly as you can."

The three wiped as much of the green Jell-O off as they could before continuing on to the next train car, which was different from any car that they had encountered thus far. It was an open-aired flat car, and it was empty except for the long sections of metal railroad track the car was hauling. Blaine wondered if they were the new sections of track that Pavel and Yuri intended to use to fix the damaged length of track up ahead at the pingo.

The group of three traversed over the sections of track to the point at which another flat car filled with more tracks preceded it. The three hopped to the second flat car and continued forward.

Accidental Algae

Blaine, Tracey, and Sveta had traversed the second flat car and were about to jump the gap to another boxcar. Suddenly, the back door burst open, and out came the crazy-eyed shoe-lace-, spider-web-, Jell-O-shooting Yuroslav Bogdanovich. Without hesitating, the mad scientist ran straight toward the three, jumping the gap and pointing the Aggrandizer.

Blaine, Tracey, and Sveta reversed course and started running back the other way. As Yuroslav chased them he put the nozzle of the Aggrandizer on the surface of the flat car and began vacuuming up small metal shavings. Tracey, who was looking back while she was running, shouted out words of warning to her two companions.

About the time they prepared to make the jump from the second flat car back to the first one, Bogdanovich began shooting. The metal shavings had been multiplied and enlarged in the Aggrandizer and were far more dangerous than shoelaces, spider webs or Jell-O. In essence, Blaine, Tracey, and Sveta were now being bombarded with shrapnel. Large pieces of metal began zipping all around the three.

Sveta jumped back to the first flat car, as did Blaine, but not poor Tracey. Just as the Sassafras girl leapt across the train cars, a large flat piece of metal hit her from behind and sent her careening down between the two train cars.

Tracey was unaware of all the things that happened next because she was squeezing her eyes shut and bracing for what she assumed would be painful impact. But rather than the feeling of pain or impact, she instead felt the feeling of . . . boogie boarding? Tracey opened her eyes and saw that she was now sliding underneath the train. The big flat piece of metal had hit her, and was now resting flat under her body and was successfully bouncing over the railroad tracks like they were waves in the ocean.

Tracey couldn't believe it; she was alive! Sparks flew every which way as she bounced along underneath the speeding train. She grabbed at anything and everything above her that she thought she could hang onto, but that proved to be difficult. She moved and sparked around almost uncontrollably.

Tracey bounced on, on her metal board, trying for all she was worth to avoid sliding or slipping off. The Sassafras girl couldn't help but think that this new way of traveling might be fun under different circumstances and in a different setting. Right now, she was just trying to survive.

At the next gap in the train, she quickly reached out and grabbed onto the coupler. She pulled herself up onto it. The flat piece of metal, now absent of its rider, violently clanked around in a shower of sparks until it finally shot free out to the side of the speeding train, landing in a small plume of dust on the tundra's cold surface.

Tracey looked up and saw that Sveta and Blaine were above her, getting ready to jump from the flat car back to the dining car. They both did so successfully, and neither of them noticed Tracey underneath, hanging onto the coupler for dear life. Wasting no time, Tracey climbed from the coupler up to the platform of the dining car, putting her back in line behind her brother and the punk rocker as they were opening the door to enter.

Blaine and Sveta stepped into the dining car, but Tracey just lay on the platform, too exhausted to move. Blaine looked back and saw his twin sister lying there. He momentarily glanced behind her, back to the flat car, and saw that Yuroslav Bogdanovich was still coming their way. "Tracey! This is no time to lie down! Yuroslav is still coming!"

Tracey realized that her brother had once more missed seeing her near-fatal fall. She gave him a weak smile, as if to say, "Just help me get into the dining car."

Blaine grabbed his sister under the arms, and virtually dragged her in by himself.

Sveta Corvette shut the door behind the twins, and then everyone just stared at each other for a moment. What were they going to do? Yuroslav was still coming after them. The dining car was still nearly filled with Jell-O. There was nowhere to go. They were stuck.

Sveta, who had basically had a smile on her face the entire time the twins had known her, suddenly had a change of countenance. The neon-green-haired punk rocker frowned. "I don't like this guy!" she stated through clenched teeth. "It is time for him to leave us alone."

Then, before they had a chance to talk her out of it, Sveta Corvette opened the dining car door and walked out to confront Yuroslav Bogdanovich. She jumped to the flat car he was on, dodging the spray of shooting metal with agility and grace. She pulled the neon-green electric guitar off her back in a ninja-like move. With a mighty swing, she knocked the Aggrandizer right out of the man's hand. The features of Yuroslav's face shifted from arrogant and evil to shocked and aghast.

Sveta didn't stop there. The neon girl picked up the Aggrandizer, vacuumed at her guitar strings, and pointed the device at its creator.

"Consider this to be your svan song!" she said to the downed Yuroslav.

The neon green rocker pulled the trigger, and within a matter of seconds, Yuroslav Bogdanovich was hopelessly tied up in oversized guitar strings.

With the threat of the mad scientist now behind them, Blaine and Tracey rushed over and ran through the remaining train cars with their friend Sveta to get to their local expert Pavel. As they quickly moved along, they saw snow for the first time here in Siberia. There were patches of it here and there in the distance. They also noticed that the mountains seemed to be getting a little bigger.

The twins had been to the arctic tundra before, while in Alaska. They had run into a lot of snow there because of a rare summer blizzard. They knew from what they had learned that in the tundra regions around the world, snow was rare in the summer time. It was typically found only in the far north. If they were seeing snow, it meant they were getting extremely far north. And, if they were getting extremely far north, they must be getting close to the place where the pingo had destroyed the train tracks. They didn't know how much more time they had until they reached the dreaded spot. What they did know was that they were feeling an urgency to make it up to the locomotive to stop the speeding train.

As if reading their thoughts, Pavel Markoff piped up on the walkie-talkie, "Blaine and Tracey, I hope you are close to reaching the locomotive, because regarding the amount of track ve have left . . . it's kind of like vhat the monkey said when he got his tail caught in the lawnmover—"

"Wait, Pavel, wait!" Blaine interrupted the engineer's riddle. "My grandpa used to tell Tracey and me this joke all the time. We are almost there, so we will tell you what the monkey said when we get to the locomotive. We are going to make it in time!"

Luckily for everyone, the Sassafras boy was right.

"It won't be long now!" Blaine and Tracey shouted the riddle's punch line in unison as they finally burst into the engine room of the locomotive.

Pavel Markoff, who was a kind-looking man, immediately gave instructions on how to use the brake to stop the train. Blaine jumped over the tied-up man, who was a bit smaller and less hairy than his friend Yuri Checkoff, to where the brake was. He followed the engineer's instructions, and brought the speeding train to a screeching halt.

The Sassafras twins and Sveta then went to work untangling the giant shoelaces off Markoff. After a few minutes, the engineer jumped up, finally free from the laces. He stretched, straightened his uniform, and thanked his rescuers.

Then keeping a promise he had made to Blaine earlier over the walkie-talkie, he hopped off the train and found a patch of lichens. "Like I vas telling Blaine earlier," the science-loving train engineer said. "Lichens are the result of a symbiotic partnership. The lichens ve are looking at here on this stone right now are the result of an alga partnering vith a fungus. The fungi protect the algae that live below, vhile the algae provide the fungi vith the various sugars they need."

The twins knelt down and looked closer at the gray curly-edged mass to see if they could tell where the fungus ended and the algae began.

"Oh my!" Pavel exclaimed. "Look at that snow patch next to us. See that bit of pink on it?"

The twins glanced up, spotted what Markoff was pointing to, and nodded.

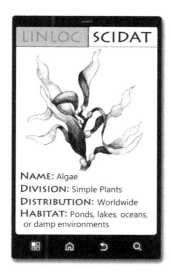

LINLOC SCIDAT

NAME: Algae
DIVISION: Simple Plants
DISTRIBUTION: Worldwide
HABITAT: Ponds, lakes, oceans, or damp environments

"That is algae, which is a simple plant," Markoff informed. "They have no roots, no leaves, and no flowers. Some species are large, like seaweed, and some are microscopic, like vhat ve are seeing now. These plants have flexible stalks called stipes and leaf-like structures called fronds. Most algae live in ponds, lakes, or the oceans. However, a few are capable of living in damp environments or on land, like this species. These algae also have a natural antifreeze that allows them to survive at these arctic temperatures. They live in the mountainous regions of the arctic tundra, just under the surface, so that they are protected from the vind. They move through the snow by using two tiny hairs, called flagella."

Blaine and Tracey opened the microscope app on their smartphones and snapped a close-up picture of the algae.

"All algae contain green pigment called chlorophyll, vhich they use to make their food through photosynthesis. But some species of algae also contain brown, red, or purple pigments." Pavel went on. "The three main types of algae—green, brown, and red, are based on their color. Red algae's red pigmentation helps them to perform photosynthesis in deep blue vater or from under the snow like our friend here. Brown algae's brown or olive hue allows them to perform photosynthesis in deeper levels of vater. Finally, green algae have an abundance of green pigment that allows them to live mainly in fresh vater."

Blaine and Tracey both let out a satisfied sigh. They were a bit chilly, but overall, they were content. They had rescued Pavel Markoff, who was yet another amazing and knowledgeable local expert. They had made friends with Sveta, the crazy-rad, very neon punk rocker. They had learned all about dwarf birch shrubs, crocuses, lichens, and algae. To top it all off, they had stopped the train before it crashed into the pingo.

As Blaine mentally reviewed their journey, he was reminded of a few lingering questions—Was the Man with No Eyebrows really still stalking them? Where had Yuri Checkoff disappeared to? Why in the world had Tracey laid down at a couple inopportune times during the train chase? Had a pingo really destroyed the tracks? The tracks looked in fine condition as far up as he could see.

Blaine was about to ask the good engineer about this last issue when suddenly a loud bone-chilling laugh pierced the cold tundra's darkening dusk. The four turned from the algae to face the sound of the laugh. There stood Yuroslav Bogdanovich. Somehow he had freed himself from the guitar strings.

The mad scientist kept laughing as he dipped the nozzle of the Aggrandizer in the snow and pushed the vacuum button. Then, before any of the four could find cover, he blasted them with an onslaught of aggrandized snow. He continued shooting until he had created a pile of snow big enough to bury forty people, much less four. He then holstered his menacing device, toned his wicked laugh down to a snicker, walked off by himself, and disappeared into the cold Siberian night.

"Would you like an ice-cold glass of lemonade, sir?" a voice asked.

Yuri reached up, rubbed his bearded face, and blinked open his eyes. Where was he? It wasn't cold—it was warm. There was no clickity-clack sound, but he was sure that he heard . . . waves?

He looked toward the voice and saw a young man in a waiter-type uniform. The man was holding out a tray covered with icy glasses of lemonade.

"Vhy, yes, I vould like one," he answered and took a glass. "Thank you very much."

"You're welcome, sir," the waiter responded. "Enjoy your time on the beach. My name is Trisno, and if you need anything else, just let me know."

The kind young man then turned and walked away. Yuri sat still for a moment and blinked just to make sure this was all real. The warm sunshine, the glowing sand, the rolling blue waves, the swaying palm trees, and the icy glass in his hand. He smiled, took a sip of lemonade, and then leaned back in his beach chair. He did not know exactly how it had happened, but he had finally made it to the beach he had been dreaming about for so long. This was exactly the kind of place he wanted to spend the rest of his life.

Chapter 12: The Mysterious Miss Été

Plant Cells in Paris

One would think that sleeping in a snow cave would be unbearably cold, but it actually wasn't so bad. It especially wasn't after Pavel had brought out some bedding from the train. That crazy Yuroslav Bogdanovich had shot piles and piles of snow at them using his Aggrandizer. He had left them there like popsicles in a freezer.

However, the Sassafras twins, Sveta, and Pavel had easily dug their way out of the snow. Then, the train engineer had quickly turned the big mound of snow into a cool, sizeable, multi-roomed snow cave for the four of them to spend the night in.

The Sassafras twins drifted off to sleep that night thinking about the plants of Siberia, speeding trains, underground science labs, and neon-green punk rockers. When they reached deep sleep, their minds seemed to fixate on one individual—Yuroslav Bogdanovich. Had he found one of the underground labs? Did those labs even really exist?

These questions eventually faded into deep restful sleep, and the next day, the twins woke up refreshed. They were back on the train riding toward Krasnoyarsk with their two friends when they finally entered all of their data and pictures from Siberia. After finding the next coordinates on the LINLOC app—longitude 2° 18' 01.3", latitude 48° 51' 05.8", they found a secluded corner on the train and zipped away.

Their destination was France. Their local expert's name was Été Plage, and their next topics of study were plant cells, photosynthesis, chestnut trees, and apple trees.

He groaned through clenched teeth that could have bitten through a three-ringed carabiner. He clenched his fists so tight that they could have crushed a smartphone to pieces. His clenched heart was angry and utterly determined to stop the twins. In Siberia, he had failed yet again. He had lunged at the twins with the Dark Cape wide open. He had intended to throw it over them and make them disappear, but he had missed.

In the end, he had been scared off the train by a truly mad scientist who had looked like a sinister version of Cecil Sassafras. Cecil was crazy in an absent-minded way, but that guy on the train had taken crazy to a whole other level. One look into his eyes would have convinced anybody of that. He had zipped back to 1108 North Pecan Street with his tail between his legs.

But no more! No more fear! No more zipping back home when things didn't play out exactly as he had envisioned. This time, no matter what he faced, be it mad scientists, pirates, darts, dogs, or Amazonian warriors, he would not zip away! He would absolutely not stop until he had the twins' phones firmly in his grasp!

They had seen pictures of it in books, but now here it was in front of them. It was towering up into the sky with more magnificent grandeur than any photograph could ever display—the Eiffel Tower. "Wow!" was the only word the twins could muster.

After zipping over the exhilarating zip lines, they had landed in customary fashion. Now they found themselves sitting alone at a small circular metal table outside some kind of café or bistro in the shadow of France's most notable landmark.

It was fairly early in the morning in Paris, so while there was sparse foot and street traffic, their surroundings were peaceful, quiet, and slow. It was a nice break from being on a speeding train running from a mad man.

A waitress came out with two menus. Before the twins even had a chance to look through them, she recommended the croissants and café au lait. Blaine and Tracey both agreed with her suggestions.

The waitress turned and headed back toward the bistro. As she re-entered, another individual came out of the building. The woman was wearing a cream-colored sundress with quite possibly the biggest hat either of the twins had ever seen. It was a huge thing with an extremely wide brim that was probably designed to shade her from the sun. It also served to completely block the woman's face from view. She walked to the other end of the patio and found

a table to sit at alone.

After a few minutes, the waitress retuned with the twelve-year olds' croissants and two cups of café au lait, which they took their time to enjoy. Blaine made sure to stick out his pinky finger while slowly consuming his delicacies so as to look cultured. Tracey just gazed at the Eiffel Tower, trying to soak up its wonder. As the Sassafrases sipped and munched, life in Paris slowly became livelier and busier right in front of them.

After a while, the French waitress returned with what the Sassafrases assumed was the check, but when Tracey opened up the miniature black folder, she found not a check but a note.

She immediately but quietly read it aloud to her brother.

You have zipped.
You have soared.
You have traveled by train.
You have picked up some science and stuck it in your brain.
If you desire to continue to do just that, walk across the patio to the lady in the hat.

THE SASSAFRAS SCIENCE ADVENTURES

The Sassafras twins looked at each other with wide eyes, and then both turned and looked across the patio toward the lady in the giant hat. Was this for real?

The last leg of their science adventures had left them more than a little wary. Then again, if Blaine and Tracey hadn't gone with the flow and moved forward with blind faith, they would have never been able to successfully progress through the given locations. So, without even needing to discuss it, the twins got up from their table and walked toward the lady in the hat.

A couple of questions went through their minds: "Was this their local expert, Été Plage? Was it the Man with No Eyebrows dressed like a lady, trying to lure them in so he could attack?" The twins were about to find out because Tracey was tapping on the woman's elbow.

Upon being tapped, the woman slowly turned her head toward the twins, the wide brim of the hat nearly hitting both of them in the face as the lady swiveled. When Blaine and Tracey saw the woman's face, they were shocked. "Summer Beach? But what? How? We thought…"

The woman was in fact their friend Summer Beach! She was obviously holding back a flood of joyful emotion as she desperately tried to retain a face that was serious, but she could do it no longer.

Her straight face gave way to a huge bursting smile as she shouted, "Blaine and Tracey Sassafras!"

She jumped up from her seat and wrapped up the twelve-year-olds in one of her patented laughing, jumping, dancing hugs. The Sassafrases were too happy to say a word, so they just let the hug happen. They knew by now that when they ran into Summer Beach, they were going to get one of these hugs.

"Did you like the mysterious poem I sent you?" Summer asked after she finally let the twins go.

"We did," Tracey answered. "It was very sneaky, but Summer,

why are you in France? How did you get here? Did you use the heliquickter?"

"No, no, I didn't use the heliquickter," Summer smiled and giggled. "Your suave and sophisticated hunk of an uncle sent me my own harness and three-ringed carabiner! Now I can travel on the invisible zip lines, too!"

The twins looked at each other and smiled. Now they weren't the only ones who were able to have the electrifying experience of zipping from place to place at the speed of light over the incredible invisible zip lines that their uncle and his prairie dog lab assistant had invented.

"Well, what did you think?" asked Blaine. "How was the zipping?"

"The zipping was immaculate!" Summer exclaimed. "The light! The speed! The rush! Oh, it was the second greatest thing that I have ever experienced!"

"What was the first?" Tracey asked.

"Meeting Cecil Sassafras for the first time, of course," the scientist gushed with dreamy eyes.

Oh, yes, how could they forget—Summer Beach, for some unexplainable reason, had a colossal crush on their eccentric uncle.

"It's strange to see you somewhere other than Alaska." Tracey grinned. "But I love that you are here in Paris with us!"

"Oh, I love it, too," Summer responded joyfully. "It is so good to get out of my underground lab and have a bit of a vacation. I have always dreamed of going to Paris, and here I am with the two cutest twins in the world!"

"But what about LINLOC, SCIDAT, and the glitch?" Blaine asked. "You know, the way that the zip line works for us is that we get longitude and latitude coordinates from the LINLOC app, but we can't move to any other location or receive any more coordinates until we enter all our required data correctly. It works like that because of some kind of glitch in the system that causes the need for the SCIDAT info to be entered before the next LINLOC coordinates can be made available."

"Everything you just explained is correct," Summer affirmed. "But that still leaves the possibility of two different modes of invisible travel. The first mode is someone like myself who has their own harness and carabiner. I wait until the two of you get all of your SCIDAT data in correctly and get your next LINLOC coordinates. Then Cecil, who records everything, gives me the coordinates, and I can zip to your location and meet you."

The twins nodded.

"The second mode that is possible for one to use the carabiner and harness would be extremely dangerous, but is possible. One could potentially flip the two coordinate rings to any random numbers, and let the carabiner snap shut. The carabiner will still do its job and connect to the selected invisible line. But with unknown numbers and coordinates, you could land absolutely anywhere. Both modes of zip-line transportation are possible without even needing a smartphone with the LINLOC and SCIDAT applications."

The twins followed Summer's line of explanation to a disturbing thought. If all of this was possible, that had to be how the Man with No Eyebrows was able to show up at so many of their locations. It was an absolute certainty that this man was traveling the globe on invisible zip lines, just like they were. They had already suspected this after seeing the man's ProLog hard hat in Uncle Cecil's basement at the end of their zoology leg, but now they were one hundred percent sure of it.

Somehow, the Man with No Eyebrows had gotten his hands on a harness and a three-ringed carabiner. He was also somehow getting their LINLOC coordinates. They didn't know how all this was possible, but it was the only rational explanation for why he kept showing up.

THE SASSAFRAS SCIENCE ADVENTURES

Chapter 12: The Mysterious Miss Été

Summer Beach broke the twins' thoughts about the mysterious Man with No Eyebrows by reaching in her purse and pulling out something new to see.

"Look, you two! Look what I brought with me to France to help you mini-frasses get your SCIDAT data for this leg of botany!"

The excited scientist now held a tablet in her hand, but this was unlike anything the Sassafras twins had ever seen. It was completely clear and see-through. The only way they could tell it wasn't just a piece of translucent plastic or glass was because its screen had images illuminated on it. The cool thing was that the images could be seen from the front and back of the tablet.

If Summer was going to help them with their data using this device, why did they need a local expert?

Blaine voiced this question, and Summer exclaimed with giggles and outstretched arms, "I am your local expert!"

The twins looked surprised.

"Été Plage is me—I am Été Plage," Summer announced, obviously overjoyed to share this little twist with the Sassafrases. "Summer in French is 'Été,' and Beach in French is 'Plage.' I, Summer Beach, am Été Plage, and I get to be your local expert for the third time!"

Full of elation and joy, Summer leaned back and shouted, "I love science! I love Sassafras! And I love France!" She sat back up and continued. "When the charming Cecil asked me to be a local expert again, I of course said, 'YES!' And when he asked me if I wanted to do the experting in France, I agreed again. And then when he told me I could travel to France using the invisible zip lines, I almost fainted with joy! So, here I am, Été Plage, in France! And I get to dive into science with you—again!"

Blaine and Tracey sat down at the round metal table, glad they got to have Summer as a local expert again. Though she was like a wound-up ball of frenetic happiness, they enjoyed her delightful antics.

Summer squealed with delight as she picked up the translucent tablet.

"Now, don't think just because I exchanged my lab coat for vacation clothes that we aren't going to get down to business here in France," Summer said in her stern voice, which was still pretty nice. "We are going to get down to business, and we are going to start right now."

The Sassafrases smiled and nodded. Summer smiled back as she ran her index finger around on the touchscreen tablet until she found what she wanted.

"The first thing we are going to talk about is plant cells," the scientist stated. "I'll read the information here on the tablet, and then, just like back in my lab, I will upload the data straight into your SCIDAT applications. So you won't have to type it all in later."

This was good news to the twins.

"I also brought along some microscope slides. After I get the information read, you two can look at those using the microscope app on your phones."

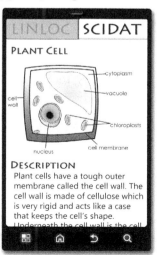

The twins glanced at each other and smiled. Summer never disappointed!

"Here we go," Summer began. "Plant cells have a tough outer membrane called the cell wall," Summer read from the tablet. "The cell wall is made of cellulose which is very rigid and acts like a case that keeps the cell's shape. Underneath the cell wall is the cell membrane."

He had successfully landed unnoticed behind some trash cans in an alley. One quick pull of the vanish string, and he was invisible. Now where were the twins? He walked to the end of the alley and peeked around the corner. Ah, there they were, sitting on an adjacent patio with that happy, pesky Summer Beach.

"The interior of the plant cell is under pressure. This turgid pressure is due to water that is present in the cell. Plant cells usually have one large vacuole which stores the cell sap, which is made up of water and glucose."

He strode quietly over to the trio. When he saw what they were doing his stomach turned in anger. At this very moment, the twins were getting the data they needed. The lady scientist was reading them information off some kind of fancy gadget. He had to stop them. He couldn't let this continue.

"Just like animal cells, plant cells have a nucleus that controls the cell activities called cytoplasm. Remember that the cytoplasm is

a gel-like substance that the organelles of the cell are found in and that the organelles carry out the activities within the cell. Plant cells, however, have two unique organelles—chloroplasts and chromoplasts. Chloroplasts contain a green pigment known as chlorophyll. This pigment gives the plant its color and makes the food. Chromoplasts contain the pigments that give flowers, fruits, and seeds their particular color."

Chloroplasts, chromoplasts, blah, blah, blah. He hated the fact that the twins were so successful at learning science this way. Cecil Sassafras was his enemy, and if he could stop Cecil's niece and nephew from learning science, it would absolutely crush him. And crushing Cecil Sassafras was exactly what he wanted to do.

It was the dark desire that drove him. After all, Cecil had wronged him in that classroom all those years ago. He could still hear the sound of all the students' laughter echoing through his mind. Maybe Cecil had done it intentionally, maybe it had been unintentional—it didn't matter. Cecil needed to pay for all the misery he had caused.

He had been relying on stealth and cunning to sneak up on the twins, but maybe that was the wrong approach. Maybe he should just tackle them right now and steal the smartphones.

"There are different types of plant cells that specialize in different functions. Two of the most common are palisade cells and spongy cells. Palisade cells are found in the upper surface of a leaf and contain lots of chloroplasts. Spongy cells have an irregular shape and are generally found inside the leaf."

Summer hit "SEND" on the tablet, put it down, and reached into her purse. After feeling around for a second or two, she pulled out several flat rectangular plastic slides and put them on the table.

"Okay, you two, grab your smartphones, open up the microscope apps, and start checking out the plant cells!" the scientist invited with her infectious smile.

The Photosyntastic Phantom Phone

There they were. The twins had just pulled the smartphones out of their backpacks. Those phones seemed to be the key to everything—the learning, the logging, the locations. He was going to do it. He was finally going to steal them.

He watched as the twins held the devices in their hands. They were opening what looked like a microscope application to take SCIDAT pictures. In the past, he would have waited for them to set the phones down somewhere before he tried to steal them—not this time. With his boldness brimming, he lunged forward.

The Sassafrases really liked the microscope application Summer had uploaded the last time they had visited her lab. The app had the ability to magnify an object up to a million times. It was fun to see the plant cells Summer had just told them about.

Blaine finished looking at the slide that had spongy cells and reached for the one that had palisade cells on it. Right about the time he got his phone over the slide to look at it, he was shoved to the ground, and his phone was snatched out of his hand!

Blaine's first thought was, "Why would Tracey do that to me?"

Tracey, who was watching her brother, had a different thought, "Why did Blaine fall to the ground, and why is his phone . . . floating?"

Blaine stood up to confront his sister. When he did, he saw that she was standing with her mouth open, pointing at something. He looked to see his phone floating in mid-air. In a millisecond, a thousand of his thoughts converged into one collective conclusion. He shouted out, "The Man with No Eyebrows!"

Yes! He had done it! He had finally stolen a phone from one of the twins! What elation! What victory! What . . . what? Wait. He was invisible, but for some reason, the phone in his hand was still visible. Maybe it was because he was holding it outside of the Dark Cape. And now look, the twins were pointing and shouting, "The Man with No Eyebrows." How had they put all the pieces together? How did they know it was him?

There was no fear in their voices. He suddenly realized they could see the phone, too. The phone must have looked like it was floating in mid-air.

Blaine dug down and found the courage that had been present when he had chased the Man with No Eyebrows in Peru after he used a robot hummingbird to spy on them. He started to run after the man once more.

He was going to get his phone back, and he was going to get

to the bottom of the Man with No Eyebrows mystery. With no fear, Blaine ran toward the floating phone, and just as he suspected, the phone started floating away from him. The chase was on.

After all he had put them through, why were these kids not more afraid of him? Why was he the one that was scared? Why did he always flee? Why was he now running off into the streets of Paris like a scared chicken?

The invisible man was being chased by Blaine, Blaine was being chased by Tracey, and Tracey was being chased by Summer. It was weird, chasing an invisible man. The only way Blaine could be positive that he was still on the man's trail was that he could see his hovering phone. It was jerking around, racing off in front of him. Blaine wondered what onlookers thought. A floating phone being chased by an American boy was probably not something that was seen every day in Paris.

They left the relatively open area around the Eiffel Tower and raced down into a maze of tightly packed streets. The Man with No Eyebrows was fast, but Blaine was not going to let his smartphone out of his sight.

The phone dodged to the left, and then to the right, and then went straight for two or three blocks before it jogged right again. The phone, then Blaine, then Tracey, and then Summer in her cream-colored dress and floppy hat raced on in a crazy-looking line.

They passed storefront after storefront, some with old architecture and some with modern style. They passed street performers, interesting sculptures, and painters working on big beautiful murals.

Paris was a gorgeous city. If the twins had been thinking about it, they would have wished they had more time to stop and enjoy it. But stopping was the last thing on their minds. Their singular focus was catching the invisible Man with No Eyebrows and retrieving the phone.

So they raced on, barely noticing any of the sights they were passing. Now the phone left the one-lane road they were on and entered a pedestrian street.

The phone weaved in and out of the plants that were there, trying to throw the chasers off its trail. It wasn't working. Blaine saw every move his phone made. He noticed he was even gaining some ground.

The plant-lined pedestrian street dead-ended at a wide ornate staircase. Without pausing, the phone left the street and bounded up the stairs. The top of the staircase met a sidewalk and then a very busy street, with several lanes of cars going each way. And, by the looks of it, the morning rush hour was now in full swing.

The phone stopped for a second, bobbed up and down nervously, and ran out into the street.

Blaine had almost had enough time to catch it! He watched his phone dodge speeding traffic with multiple near misses. The drivers of the cars couldn't see the invisible man and probably couldn't see the phone, either, so they were not swerving to miss it.

Blaine looked to his left and saw a pedestrian bridge arching over the busy street. This would be a much better way to cross the road. He raced to his left, and then hopped up a winding set of stairs two at a time, before crossing the bridge. He kept his eyes on his phone the entire time. Somehow, it made its way across the busy street without getting smashed. It then took a right on the sidewalk and raced away from Blaine and the pedestrian bridge.

The twelve-year-old was able to keep the handheld device in his sights. He made it safely down the pedestrian bridge's other set of stairs and continued the chase, with his sister and Summer not too far behind him.

The phone raced straight down the sidewalk until the sidewalk was interrupted by the wide driveway of a big fancy hotel. Blaine's phone ran up the driveway, past a gurgling fountain, and under a huge ornate awning, where several valets were in the process of taking keys from incoming hotel guests.

The smartphone swung up, hit one of the valets right under the chin, and grabbed a set of keys out of the reeling man's hand. They floated in the air, just like the phone, and then both mid-air objects got into the waiting car. The door closed, the engine started, the tires churned, and then the sleek, red sports car sped down the driveway toward the busy street.

"No!" Blaine shouted as he came to a stop, watching the sports car race past him. He had not caught the Man with No Eyebrows. He had lost his phone.

The boy's shoulders slouched, but just as they did, his sister and Summer Beach raced past him.

"The chase isn't over yet," Summer shouted optimistically. "C'mon, Blaine, let's go!"

The scientist ran up to one of the other valets and grabbed the car keys out of his hands.

"Merci beaucoup!" she sang with a smile as she loaded herself and the twins into a sparkling green-blue car, which resembled the one the Man with No Eyebrows had taken except that it was a convertible.

"I promise we will bring it back!" Summer shouted pleasantly to the stunned valet as she shut the car door and cranked the engine to life.

"Hold on, Sassys!" she said a split second after her foot had already hit the gas pedal. Été Plage left black tire streaks all the way down the driveway and into the busy street as the chase changed from a foot race to a car chase.

Upon reaching the road, the three immediately spotted the Man with No Eyebrow's car a few hundred feet up in front of them. At first, it was slow going within the busy flow of traffic, but that quickly changed. As soon as the invisible man driving the car saw them pull into the street behind him, he pushed the gas pedal down. He started weaving in and out of traffic like a man full of desperation.

He sped and swerved around the other cars on the road as if they were standing still. With her left hand on the steering wheel, her right hand on the stick shift, and her feet on the pedals, Summer began maneuvering the car through the Paris traffic like a pro.

Suddenly, the Man with No Eyebrows yanked the wheel of his red car, sending it screeching off onto a small side road. Summer followed, taking the sharp turn with no problem. The three now found themselves chasing the Man with No Eyebrows down streets in their car that were similar to the small tight ones they had just chased him down on foot. The Sassafras twins held their breath as the speeding, roaring sports cars flew down through the narrow streets. They tore around blind corners and went the wrong way on one-way streets. Blaine and Tracey gripped the car's leather seats with white knuckles. Summer Beach, however, looked as cool as a cucumber.

"Well, while the three of us are all in the car together, why don't we go ahead and talk about photosynthesis," she suggested with the peace and calm of someone who was flying a kite. "Does that sound good to you two?"

The twins would have answered, but they couldn't get their voices past their hearts, which were currently lodged in their throats.

"Well, okay then!" Summer acknowledged as if Blaine and Tracey had answered with a resounding yes. "Photosynthesis literally means 'putting together through light.' It is the process plants use to get the food and energy they need from the sun."

They made a hard turn to the left, into an alley, narrowly missing a dumpster.

"In photosynthesis, the plant absorbs water from the soil through the roots and carbon dioxide from the air. Then light, which is absorbed through the leaves, is used to convert the water and carbon dioxide into glucose and oxygen."

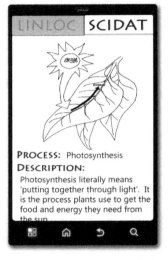

There was a hairpin turn around an oblong building—tires peeled as they turned into the skid.

"The plant releases the oxygen as a waste product and uses the glucose as food. The plant can also use the glucose to form other substances they need, such as the cellulose that is used to form the cell wall and starch, which acts as a food store in the seed."

A delivery truck blocked the road, requiring a slight swerve up onto the sidewalk as they zipped past lampposts. Then, another swerve back on the street.

"Photosynthesis takes place in the chloroplasts found in the plant cells. The chloroplasts contain stacks of membranes coated with chlorophyll that act like solar panels to capture energy from the sun's light."

Up ahead, was there a drop-off? No, it was a staircase leading down. They weren't stopping. They were airborne. They landed with sparks, just behind the red car.

"A plant's leaves are flat so they can absorb more energy from the sun," Summer said as they burst out of a skinny street back onto a wide street, swerving and spinning to miss cars coming from both directions.

"Plants are not the only living things that use photosynthesis. Many bacteria also use this method for producing food."

The wide street was getting even wider, and the traffic began thinning out. They could see now that they were on the road leading out of the city. Finally, the twins were able to breathe again.

"That, my young Sassafrases, is the lowdown on photosynthesis. When we get Blaine's phone back, I will upload all of that information for you. The best pictures you are going to find on photosynthesis are probably going to be in the archive app. So, that is another thing we will need to do once we get this runaway smartphone back. Oh, look! The invisible man is driving out of the city on the highway. Now we will really see what this car can do!"

The Sassafrases lost their breath again as they flew down the highway.

Chapter 13: A Tour of Versailles

Crashing Chestnuts

The Sassafras twins thought back through the villains they had faced while zip-lining around the world studying science. There was Yuroslav Bogdanovich, the mad scientist in Siberia who looked like a sinister version of their uncle. There was Rama, the pirate king, who had chased them through the jungles of Borneo. Who could forget Manuel Hernandez, the rich fox-napper and all-around cheat who thought he reigned supreme on the Argentinian pampas?

There was Kingman Nowarak, kidnapper of Thai women and children. Smirk and Chili, who were the high school bullies in Lubbock, Texas. Dr. Veeginburger was not really a doctor, but a dealer of illegal supplements. Salvatore and Bruno were the unoriginal chefs from Venice. There was Ortiz, the rogue logger bent on taking down every tree in the Amazon rainforest. Tank and Billy Smith were the mean Canadian farm boys, who had tried to prank the twins out of their jobs. And, finally, the dark-robed, sword-wielding Itja leader of the Kekeway bandits.

Some of the villains had been brutes. Some had been tricky. All of them had been, on some level, imposing. But by far, the most consistent evil-doer had been the Man with No Eyebrows. He had first shown up in Kenya—the twins' inaugural trip on their zoology leg. He had dogged their adventures after that, and here he was now, in France.

The Sassafrases had the sneaking suspicion that he was often present, even when they didn't see him. After all, he had also somehow stolen Phil Earp's Dark Cape suit, which gave him the ability to be invisibly stealthy. Blaine and Tracey had no idea why this mysterious man followed them, but they were ready to put an end to it.

CHAPTER 13: A TOUR OF VERSAILLES

The greenish-blue sports car the twins were riding in with Summer Beach had long since been in its highest gear. The car cruised at high speeds on a highway leading out of Paris as they chased the red sports car being driven by the Man with No Eyebrows, who had stolen Blaine's smartphone.

With skillful and precise driving, Summer had managed to edge closer and closer to the red car, but the Man with No Eyebrows still had a lead. He also still had invisibility thanks to the magic Dark Cape suit.

Blaine and Tracey could not get over how strange it was to see a car that looked like it was being driven by no one.

"It looks like he is heading into Versailles," Summer said, still maintaining her relaxed demeanor.

"It is a little greener out here than it was in Paris, is it not?" the scientist mused as she scanned their changing surroundings.

The twins nodded.

"Versailles is said to have some of the most beautiful gardens in the world, and check out these trees that line the highway!"

Summer remarked.

Though they were still a little nervous to be going this fast in a car, the Sassafrases looked around and tried to enjoy the scenery. There were indeed beautiful trees on each side of the highway. There were also many chateaus with well-manicured gardens in the distance.

"This part of France has many of the very same trees and plants that are found in deciduous forests around the world. These forests are dominated by broad-leaved deciduous trees such as the oak, birch, maple, and beech. Since most of the trees lose their leaves in the winter, the forests go through some big changes throughout the seasons.

"The deciduous forest is typically found in moderate climates—areas with warm summers, cool winters, and moderate precipitation throughout the year. There are usually two or three layers in the forest. First, there is the canopy of the trees at the top. Then, the ferns and small plants on the ground. And, finally, the layer of shrubs in between."

Just as the scientist finished her sentence about the layers of the forest, the Man with No Eyebrows swerved his car from the far left lane all the way over to the right and shot down a small exit ramp, leaving the highway.

The twins guessed that he thought a sudden and dangerous move would help him lose his pursuers, but it did not. Summer executed the same move with their car, and they ended up even closer to his car.

The exit ramp ended in a "T," leaving the choice of going left or right. The Man with No Eyebrows went right, taking the corner quick and speeding forward. Summer and the twins followed close behind. They now found themselves on a winding wooded road speeding through the countryside of France.

Both sports cars took the curves of the country road with

grace as they were designed to do, but it was apparent that Summer was the better driver. She was driving this road as if she had driven it a hundred times before. Whereas the Man with No Eyebrows was driving it like every turn was a new surprise to him.

A sharp left, a tight right, a series of zigzags, a straightaway between the plots of two huge chateaus, then into more precarious curves. Then, they hit a wide sweeping right, ending with an abrupt left. This was a tricky turn, and it proved to be too much for the Man with No Eyebrows and his red sports car. His tires screeched, and his brake lights shone as he spun off the road into a cluster of trees.

Summer brought the vehicle to a stop and jumped out of the car first, followed by the twins. All three quickly ran down to where the Man with No Eyebrows had come to a stop.

"Oh, my!" Summer exclaimed when she saw exactly what had happened. "He crashed it pretty well, didn't he? Right into that chestnut tree!"

The Sassafrases saw what she was talking about. The red sports car now had a crumpled front end. Their hearts were pounding, which had nothing to do with the crash. They were anticipating a confrontation with the invisible villain who had literally chased them around the world.

The three watched as the driver's side door of the red car opened. There was the sound of some labored movement as Blaine's phone appeared. It was still floating in mid-air because it was still being held by the invisible man. Blaine summoned his courage and lunged at his phone and the invisible form that was holding it. There was a collision as the adrenaline-filled boy hit the full-grown invisible man. Blaine managed to knock the man over, but the phone remained firmly in the villain's grasp.

He kept reaching for it, but the Man with No Eyebrows kept moving his arm around wildly, managing to keep the sleek device away from its owner. The two rolled around on the ground in a crazy, clawing wrestling match. Finally, the Man with No Eyebrows managed to get to his feet, but the determined twelve-year-old, Blaine, was hanging on to his back. Blaine had never experienced anything so strange. He could feel the man punching and swiping and writhing as well as the black suit the man was wearing, but he could see nothing.

It was insane to Blaine, but it looked even crazier from the vantage point of Tracey and Summer. All they could see was Blaine floating around in the air all by himself—grabbing and grasping and reaching for the floating phone. The boy looked like he was riding an invisible mechanical bull.

Blaine kept his left arm tightly locked around the man's neck and kept reaching for the elusive phone with his right hand, but he couldn't quite get to it. Suddenly, the boy felt a big gloved hand clutch the back of his neck. The next thing he knew, he was flipped over frontward and landed on his back on the ground. The Man with No Eyebrows had flung him off. Blaine then felt something being draped around his body. That was not a good thing.

Tracey shook off her frozen fear and resolved to run in and help Blaine. She grabbed Summer's hand and started to run. Just as Tracey decided to do that, Blaine was flipped forward onto the ground, landing hard on his back. Then right before the girl's eyes, the worst possible thing happened—Blaine disappeared.

Blaine could tell that his body was being transported between places, but it was nothing like traveling on the invisible zip lines. There was no light, no rush, and no exhilaration. Just darkness and the slight awareness that his body was in motion. Then, abruptly, there was light . . . and sand . . . and the sound of water. Blaine immediately knew he was not alone. He knew that the Man with No Eyebrows was there with him.

Blaine spotted his phone in the sand. He reached for it, but he was stopped by the invisible familiar form. The boy's grasping hand felt a cape and then a string. He pulled the string, hoping it was the vanish string. The Man with No Eyebrows suddenly appeared in front of Blaine, dressed in the imposing Dark Cape!

The boy took a step back and looked at the man, who was clenching his fists and glaring at him from behind the black-tinted visor of the Dark Cape's helmet. A few shakes of fear peppered Blaine's resolve but not enough to curb his determination. He glanced down and saw that his phone was still lying in the sand at the man's feet. Not wanting to wait for a better opportunity that might or might not present itself, the Sassafras boy made a lightning-fast dive for the phone.

He got it! He smiled, but now what was he supposed to do? Blaine started to run away, but the Dark Cape tackled him from behind. The two went careening over the sand and down into the water. Blaine lost the grip he had on the phone, and he watched as it went flying out into deeper water. It landed with a splash and then disappeared.

"No!" Blaine shouted.

The Dark Cape wrapped him up in a vice-grip bear hug and started to pull him down under the water. In borderline panic, Blaine grasped and clawed at the first thing his hands found—the Dark Cape's helmet. Now it was the Man with No Eyebrows who was visibly flustered. He quickly let go of Blaine, turned around, reached for the vanish string, and disappeared.

The water around Blaine became relatively still. He stood breathlessly for a moment and then let out a long relieved sigh. He was alone. The Dark Cape, the Man with No Eyebrows, was gone. But the boy's relief lasted only for a second. "My phone," he thought. "I have to find my phone!"

Tracey had never seen Summer Beach like this. The female scientist was actually nervous. When Blaine had disappeared, Tracey had suggested that they call Uncle Cecil to tell him what had happened and to see if he had any ideas about what to do. Summer had agreed and had been the one to make the call. Now that she was actually talking to Cecil Sassafras, she could hardly get a word out—much less a sentence.

The poor love-struck scientist gave up trying and handed the phone to Tracey, "Hello . . . Uncle Cecil. Did you gather what happened from what Summer just said?"

"I sure did, Blaisey," Cecil responded with deflated enthusiasm. "Train has disappeared, and it is all because of that Man with No Eyebrows."

"That's right," Tracey replied, amazed that Uncle Cecil had been able to translate Summer's love-struck mumbling.

"What should we do, Uncle Cecil? How can we find out what happened to Blaine?" the girl asked, alarmed.

"Don't worry, Blaisey," Cecil commiserated. "We will find Train. What you need to do is finish logging the SCIDAT data for your current location and be ready to zip to your next destination. I am still tracking your brother's smartphone location on the big screen. It looks like he is currently in . . ." Cecil paused as he looked at the big world map with Blaine and Tracey's two lights.

". . . Borneo. For some reason, he seems to be back in Borneo. It's not good that he's back there, but at least we can still track him. Don't worry, I'm sure your brother will call me just as soon as he can. We will figure this out!"

There it was! He had just stepped on it with his right foot. Blaine reached down and pulled his smartphone out of the water. The boy raised his arms in victory, but his elation over finally finding his phone was short-lived. He realized that the water-logged device probably would not work. Blaine let all of the water run out of his phone. He reluctantly placed the device in his pocket and sighed.

His first move would have been to call Uncle Cecil, but that was now impossible. What was he going to do? Blaine scanned the beach from his spot in waist-deep water. Where was he? There were beach chairs and palm trees and a sandy ridge. Was he back in Borneo, at Pitchers Beachside Resort? If he was, he could get his friends, Trisno and Novi, and see if they could help him.

But what about the Man with No Eyebrows? What if he came back? What if he had not really left? He would be watching Blaine now, ready to pounce. Blaine's adrenaline had already evaporated, and a lonely fear began to set in. Suddenly, his only desire was to be reunited with Tracey and Summer.

Blaine's mind raced. How could he get out of here? There had to be a solution. Wait—what was it that Summer had said about possible modes of invisible zip-line travel without a smartphone?

"Chestnut trees are broadleaf trees in the same family as the oak or beech tree," Summer read from the translucent tablet.

Tracey had filled the scientist in on what Uncle Cecil had told her their plan of action should be. So, Summer was now wholeheartedly going along with that plan by sharing with Tracey the information she needed about the chestnut tree.

"Trees can be either deciduous or evergreen." Summer continued. "Deciduous means that they shed or lose their leaves in certain seasons. Evergreen means that their leaves stay green year round. This chestnut tree is deciduous. It has two types of flowers—the male and the female. The male flower matures first and releases its pollen, which is carried by the wind or by insects to the female flowers.

"The pollen needs to reach a different chestnut tree because they cannot pollinate themselves. This is known as cross-pollination. Once pollinated, the female flowers develop into fruit, which matures in the fall. The fruit of the chestnut tree is a prickly ball, called a bur, but when you open it up, it reveals several mahogany-colored seeds. The seeds, or nuts, have a creamy white flesh that turns sweet when roasted."

Tracey used her phone to snap a picture of the chestnut tree in front of them. She was very careful not to get any part of the smashed-up red sports car in her photo. Though it was summertime, and she was feeling gloomy over the disappearance of her brother, she started to hum a certain familiar Christmas carol.

Summer hummed along with Tracey for a bit before they entered into a rendition of the song with giggles. Summer then read more about the tree. "The chestnut tree is nicknamed the 'bread tree,' because during medieval times, chestnuts were used as a main source of food for many people. The nuts contain twice the amount of starch as the typical potato and they are the only nuts known to be a source of vitamin C. These trees can live for centuries and grow to be over ninety feet tall." The three-time local expert stopped and uploaded the information to Tracey's phone.

She then added cheerfully, "The palace garden, in Versailles, has over eighteen thousand of these beautiful chestnut trees. It looks like we are right on the edge of the palace's property."

Tracey smiled, trying to be as cheerful as Summer, but she was really worried about her brother.

Summer sensed this and said sympathetically, "We will find him, Tracey, don't worry. In the meantime, let's do all we can to be ready for that reunion when it happens. We can use our nervous energy to hike down this trail and go find some apple trees!"

The Apple Festival

He usually did not know where exactly he was going to land at each destination. This time he didn't exactly know where on earth he was going to land. Blaine had turned the two coordinate rings on his three-ringed carabiner to random numbers and had let it snap shut. Just as Summer had said it would, his carabiner had connected to the invisible line. Blaine had hung, waiting for a few seconds, with his feet dangling in the water. He was wondering if he had made the right decision. Then, zip, he was off!

With no SCIDAT, no LINLOC, no working phone, and no twin sister, the Sassafras boy flew through swirls of light. In a matter of seconds, he made his landing with a tingling body sapped of sight and strength. When his faculties returned, where would he be?

Blaine found out soon enough, and it wasn't good. Apparently, he had landed on top of a volcano. More specifically, he was inside the top ring, on a steep pitch that sloped downward and led to a cauldron-like circle of glowing orange lava. His body was on a pile of loose shale, and he was sliding down toward a hot orange glow.

"Not good!" Blaine screeched as he frantically fumbled around with his carabiner, trying to quickly set new numbers on the coordinate rings.

He was already too close to the circle of fire.

"New longitude coordinate . . . done," he stated out loud to no one in particular.

He was picking up speed. "New latitude coordinate . . . got it!"

Blaine was just feet away from falling into the lava.

"Carabiner snapped shut . . . and locked." He rejoiced, but was it too late? Blaine could feel himself falling into the fiery volcano. Then, suddenly, abruptly, joyfully, he was yanked up. He was hanging over the boiling, steaming volcano attached by the harness and carabiner to an invisible zip line. He sweated profusely as the seconds until takeoff ticked by, and then he was off again.

He threw down the helmet in disgust and watched as it bounced a couple of times. It landed on the suit and cape on the ground in front of him. He had had enough of this despicable Dark Cape suit. It just wasn't working. Yes, it could make things disappear. Yes, it enabled him to become invisible, but had it helped him stop the Sassafrases? No!

He sulked down into a chair and furrowed his hairless brow. The Dark Cape suit was a bust. The expandable trap boxes had been a bust. The robot hummingbird and arctic ground squirrels had both been busts. Countless other stealthy plans that he had made and tried to execute had failed. The twins just seemed to be unstoppable.

The past actions of that silly Cecil Sassafras remained

unavenged. For the first time since that terrible incident in that junior high classroom, he thought about quitting. He sat there and seriously considered putting a stop to his run on this track toward revenge.

No! What was he thinking? He couldn't quit! He had to get back at Cecil Sassafras and make him pay! It had become his life's goal. He had tucked away several more ideas how to stop the twins from learning science. One idea in particular stood out, but it needed more work.

New plan—he would leave the twins alone for the rest of their study about botany. This would give him time to put the finishing touches on this next idea. Then, he would come after them again in full force at the start of their next science-learning leg!

As Blaine's senses normalized, he was glad that he felt no heat. He relaxed a bit, knowing he had not landed on another volcano. As his senses returned, he realized it was quite the opposite. All he felt was extreme cold and a flat surface. Blaine attempted to stand up, but just as he did, he slipped back down to his seat. He must be on a solid slab of ice.

The Sassafras boy looked around in every direction. He couldn't see much because there was quite a bit of snow in the windy air, but it was pretty clear that he was sitting on the surface of a frozen lake.

Blaine made another attempt to stand to his feet, and this time, he was partially successful. He got to his feet, but he slipped and slid all over the place. He found himself moving forward in a half-run, half-walk on the ice. He made it several yards before he fell again. This time, he fell fairly hard. When his body hit the ice, a crack about as long as he was immediately appeared.

Blaine lay there, completely still, staring at the crack. He did not want to fall through the ice into a frozen lake. The crack momentarily remained as it was, but then all at once, it exploded into a spider web of cracks that branched out in every direction.

"Not good!" Blaine screeched again. He quickly reached for his carabiner, grabbed it, and immediately spun the coordinate rings to new numbers.

The sound of cracking ice intensified, and Blaine knew if he stayed on the spot he was on, he would fall through at any moment. Even though his fingers were freezing, the Sassafras boy managed to turn the rings to new longitude and latitude coordinates in record time.

The carabiner snapped shut.

The ice broke.

Blaine's legs dipped into the freezing water.

Then, he was pulled up by his harness as the three-ringed carabiner found its next zip line. Blaine hung in the cold air for a few seconds with his fingers crossed, hoping that the third time would be the charm.

In a flash, he was gone.

Versailles, France, was one of the most beautiful places Tracey Sassafras had ever seen. As she and Summer walked the wide gravel path in the woods, Tracey couldn't help but think that Versailles was like one of the enchanted lands found only in fairy tales. It was absolutely breathtaking.

Summer was walking at a gentle pace, and for that, Tracey was glad. This speed allowed her the chance to soak up the gorgeous scenery more easily.

"Oh, look! Look, Tracey!" Summer Beach said joyfully. "Look at all the apple trees! We've found an apple orchard!"

Although not completely ideal, this location was much better than a fiery volcano or an icy lake. Blaine's third set of randomly spun coordinates had put him on an island. An extremely small island. It was an almost perfectly circular sandy island, about ten feet in circumference, with a lone palm tree protruding up from the very center.

There was no sight of land anywhere else, just calm blue sea in every direction. It was a rather peaceful place, but Blaine was definitely not at rest. He needed to get back to his sister, and this random calibration of coordinates just wasn't working.

The boy took off his backpack and sat down, leaning up against the palm tree. He needed to slow down and think. He couldn't keep doing the kind of zipping he had been doing.

He let out a long deep breath and then took in a full breath of the salty sea air. What was the best plan of action? His mind drifted back to when Summer had explained the two ways that one could travel on the invisible zip lines. He remembered she had said they could travel with random or specific coordinates. The random coordinate mode of traveling was available at all times, but it was clearly unpredictable and could be dangerous. Blaine could vouch for the truth of that!

The specific coordinate mode of traveling was much more precise, but it was offered only after correct SCIDAT data had been entered. Uncle Cecil recorded all of the specific coordinates, and they could be used after the fact by anyone with a special three-ringed carabiner, like the people Cecil gave them to or the one who had somehow stolen one.

The question for Blaine now was whether he had recorded any of the previous coordinates in his mind. More precisely, he needed to remember the coordinates for Paris, France. If he could recall those, he could zip back to that location before Tracey moved to the next location, and they could be reunited.

Blaine began mumbling coordinate numbers out loud in an attempt to remember the most recent numbers for Paris. "Longitude 18° 5' 04.4", latitude 51° 6' 03.1", was that it? No, no, maybe it was longitude 2° 48' 02.1", latitude 56° 21' 08.8". No, I am sure that's not it, either. I know, it was longitude 101° 65' 11.2", latitude 36° 14' 0.1"? No, no, no—I am never going to remember what the numbers were!"

Blaine threw his hands down to the sand in disgust. Just as he did, a coconut fell out of the palm tree and hit him right on the head.

"Longitude 2° 18' 01.3", latitude 48° 51' 05.8"," Blaine said confidently, a welt immediately forming on his head. "Those are the coordinates for Paris, France! I am sure of it!"

Tracey and Summer had found the apple orchard. And to their surprise and joy, they saw that something called the Versailles Summer Apple Festival was in full swing. The two exited the woods and walked through the many festival booths.

Almost everything that one could think of to do with apples was being displayed there. They saw people apple bobbing, apple bowling, apple throwing, apple catching, and apple shooting. There was apple carving and apple juggling. There were people dressed up as apples and people with apple face paint. There was a makeshift outdoor theater, where a video about apples was playing. In addition, there was a seemingly endless variety of apple-flavored treats and drinks available.

"Oh, look, Tracey!" Summer Beach pointed, clapping her hands as she did a mini-dance. "There is an apple tractor ride!"

"An apple tractor ride?" the Sassafras girl questioned.

"Yes, it's where they slowly drive you through the apple orchard on a flat-bed trailer being pulled behind a tractor. Oh, let's do it, Tracey! Let's ride through the apple orchard!"

A few moments later, Tracey was sitting next to her overjoyed local expert on a slow-moving trailer, rolling through the beautiful apple orchard. Summer had already pulled out her see-through tablet and was providing a commentary for the ride as she read the information she had found on apple trees.

"The apple tree is another deciduous broadleaf tree. Like all other broadleaf trees, they have a singular woody trunk, which thickens as the tree grows older. From the trunk, the tree divides into branches, which form the crown of the tree. The crown holds the twigs, leaves, flowers, and fruit. Apple trees have been cultivated for thousands of years because of their sweet edible fruit.

"The fruit matures in the fall and has a sweet white to yellow flesh that surrounds the seeds in the center. There are typically five seed pods in the center of the apple, arranged in a star-shaped pattern. The outside of the fruit is covered by a tough outer skin that can range in color from green to red to yellow."

Tracey gazed through the rows and rows of apple trees in the orchard. She could see a virtual rainbow of apple colors on display before her.

"Like other trees, the apple tree has long woody roots." The scientist continued. "The roots spread out in the soil to absorb the water and nutrients that the tree needs. A special kind of fungi grows on the roots of apple and other broadleaf trees. It helps to supply the tree with nitrogen and other minerals that it needs to grow. Overall, the typical apple tree grows around nine to fifteen feet tall. Pollination of the apple tree occurs in the same way that it does in the chestnut tree, except that the male and female flowers of the apple tree bloom around the same time."

"You mean the pollen from the male flower is carried by the wind or insects to the female flower on a different tree?" Tracey asked as she recalled some of the details about pollination of the chestnut tree.

"Exactly!" Summer replied, impressed with how quickly Tracey was catching on.

The Sassafras girl snapped a good picture of an apple tree as the tractor ride came to an end.

Tracey looked around as she stepped off the tractor. This place was beautiful. In one direction, she saw neat line after line of fruit trees. In another, she saw rows of well-maintained vegetables and herbs.

Summer pointed toward the garden as she said, "That is the 'Kitchen Garden of the King' or 'Le Potager du Roi' in French. It was started under the command of Louis the Fourteenth to provide food for his court. Many say it is the most extensive garden ever grown. If you look off in the distance, you can catch a glimpse of the palatial palace of this king."

Tracey sighed in amazement. This was quickly becoming one of her favorite spots. She just wished Blaine could have experienced it, too. As the two girls drank in the sights, the twelve-year-old's phone rang—something it had never done.

"Hello?" Tracey answered.

"Blaisey. It's Uncle Ceci-frass! I have some great news!"

Chapter 14: California Creatures of Mystery

Squatchin' with the Sitkas

Uncle Cecil's news was great! Over the phone, he had informed Tracey that, according to the tracking screen, Blaine had managed to zip back to France. Tracey and Summer had hurriedly returned to the car and zoomed back to Paris. They returned the sports car to the valet, whom they had borrowed it from, and ran from the big fancy hotel to the café patio near the Eiffel Tower where the Sassafras twins had landed.

The tracking screen had been correct because Blaine was sitting there, sipping another café au lait with a wry smile on his face. The girls listened as Blaine recapped his story about what it had been like to wrestle an invisible man and travel on the invisible zip lines with random coordinates. Tracey told him about the much less harrowing time they had had at the Versailles Summer Apple Festival. Summer suggested that Blaine try to turn his phone on just to see if it might work. To their surprise, it re-illuminated, and all of its capabilities were intact even though it was not still completely dry. The smartphones Uncle Cecil had given them were not only sleek; they were resilient, too.

Summer re-uploaded the information about chestnut and apple trees to Blaine's phone. The twins then sent that information with pictures to Uncle Cecil's basement. Blaine sighed and laughed happily as he looked at the next coordinates on his phone via the LINLOC app. It read longitude 124° 13' 42.9" W, latitude 40° 10' 14.6" N. Even though some of their past landings had been precarious, this was the way to travel—invisible zip lines with specific coordinates that would land them in precise locations. Blaine hoped he would never have to travel with random coordinates again.

"Our next local expert's name is Brock Hoverbreck," Tracey read the LINLOC information aloud. "We'll be going to California, and our topics for study will be Sitka spruces, conifer cones, coastal redwoods, and mushrooms."

"No volcanoes, no frozen lakes—California, here we come!" exclaimed Blaine.

"What about you, Summer?" Tracey asked. "Are you going to come with us?"

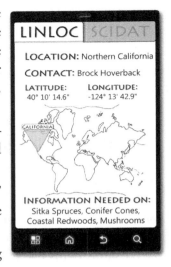

Summer seemed to be frowning and smiling at the same time.

"No, little Twinkle-frasses. I won't be going to California with you. I will be headed back to the lab in Alaska to keep plugging away at various projects—after a few more days of vacation here in beautiful France, of course!"

The scientist's face was now showing a huge smile. "I am so proud of you two! Keep going! Don't quit! Don't worry about that Man with No Eyebrows! Just keep learning! You never know when you might see yours truly again. I am rooting for you!"

A happy jumping dance-hug ensued. Goodbyes were spoken, and then, the twins zipped their way to California.

The landing was—not sandy-beachy, more like carpety-beachy. The twins could hear the sound of ocean waves and smell salty sea air, but they were pretty sure they were lying on carpet. How was that possible?

When the ability to see had fully returned, the Sassafrases saw that indeed they had landed on thick purple shag carpet. The carpet was on the inside of a van, and the van was parked at the beach, which explained the sounds and smells of the ocean.

The van was filled with more than just shag carpet . . . much more. Much of the interior complemented with the dated carpet. The twins saw hanging wooden beads, posters of sixties and seventies musical heroes, glass vases containing daisies and sunflowers, black lights, peace symbols, velvet curtains, and even a small disco ball fastened to the ceiling. The van was also equipped with a small kitchenette that looked as though it had recently been used. Most curious of all was the rack of metal shelves affixed near the back doors. The rack was piled high with strange gadgets and equipment.

After spending a few minutes staring at their weird new landing location, the twins pulled on the door handle and slid the big side door of the van open. They stepped out into the sand and saw they were on a very long beach, bordered on one side by the ocean; the other side was a towering forest. They turned around and caught their first full glimpse of the van.

They were amazed to see that the outside was decorated just as elaborately as the inside, if not more. It was a Volkswagen that was hand-painted with the vibrant colors of the rainbow arching and swirling all over. Set inside the rainbow scenes were images of unicorns, elves, sea serpents, aliens, and various other mythical creatures. Then in big bubble letters it said—C.O.M. Carriage.

"C.O.M. Carriage?" Blaine exclaimed full of curiosity. "What in the name of Rutherford B. Hayes does that mean?"

"I don't know," Tracey answered. "But maybe these people walking our way will be able to tell us."

Blaine looked in the direction that his sister was looking and saw that six people were walking up the beach toward them. The approaching group was made up of four men and two women. They were laughing and walking slowly. Both of the women and three of the men were wearing tie-dyed T-shirts. The additional man was wearing a camouflage T-shirt. All the shirts had "C.O.M. Crew" on the fronts.

"Very curious," the twins thought. C.O.M. Carriage, C.O.M. Crew—they had no idea what was going on, but the letters C.O.M. seemed to be of some importance. The group of six sauntered up the beach toward the van. It wasn't until they were practically upon the vehicle that they noticed the twins standing there.

"Well, hello there, little tikes," one of the men who had long hair and orange-tinted sunglasses said. "My name is Ned. Who might the two of you be?"

"We are Blaine and Tracey," Blaine responded. "Blaine and Tracey Sassafras."

"Groovy." The long-haired man nodded happily and slowly.

The man with the camo T-shirt spoke up with much less cheer in his voice than Ned. "You two here to go squatchin'?" He used one hand to groom his short mustache and the other to brush through his short crew cut.

"Squatchin'?" Blaine asked, puzzled. "What's that?"

"You don't know what that means?" another one of the men

gasped. He was the tallest of the group.

He then turned to one of the women and poked her in the shoulder a couple of times. "Tell 'em sis. Tell 'em what squatchin' means."

The woman, who was red-headed and very pretty and kind looking, chuckled at her poker, who was presumably her brother. "Slow down, Cory. Before we jump into that, let's see if they are hungry. Harmony, do we have any of the food from our picnic left over?"

She turned to look at the other woman, who looked very similar to her except that she had blonde hair. The woman named Harmony smiled and began rummaging through a bag slung over her shoulder.

"We sure do, Melody," Harmony responded. "There are a couple of sandwiches left. Would you like them, Blaine and Tracey?"

The twins, who rarely ever turned down food, nodded. They took the sandwiches from Harmony and immediately began munching. The man who hadn't spoken yet was a good-looking guy with slightly spiky dark hair. He grabbed some bean bag chairs out of the van and set them out on the beach, creating a spot for everyone to sit down.

When the group of eight had done so, Melody spoke up again.

"Before I tell you two what squatchin' means, let me introduce everyone," the pretty redhead said, addressing the Sassafrases. "My name is Melody Albermully, and I serve as a sort of manager to this ragtag group of hippies you now find yourselves with."

Everyone smiled at being called a "ragtag group of hippies" as though it was a compliment.

"That's my sister," Melody said, putting her hand on the blonde-headed woman's shoulder.

"Her name is Harmony, and she is our historian. Next to her is Rip," Melody said, speaking of the man wearing the camo shirt. "He is our tracker."

"Hunter," Rip corrected.

"Tracker," Melody asserted nicely but firmly.

"After Rip is Sam." Melody continued, introducing the man with spiky dark brown hair. "He is our tech specialist."

Sam gave a slight wave to the twins. They waved back.

"Sitting next to Sam is my younger brother. He's the runt of the Albermully family at nearly seven feet tall! His name is Chorus, but he goes by Cory. He serves as first technical assistant to Sam."

"Wow, almost seven feet tall," Blaine thought, a little jealous.

"And last but not least is Ned." Melody pointed toward the long-haired sunglass-wearing man. "He is our driver."

Ned bobbed his head and smiled, flashing a peace sign toward the twins.

"Now onto the definition of squatchin'," Melody continued. "It means to go looking for Bigfoot."

"Bigfoot?" the twins exclaimed.

"Yep," Melody replied. "At least that is what he is called in most of North America. Bigfoot is a Sasquatch, so when an individual or a group goes out looking for him, we call it squatchin'."

The Sassafrases now grasped the term squatchin', but they were still wondering about the C.O.M. lettering. Tracey voiced their curiosity. "We know what the words crew and carriage mean, but what do the letters C.O.M. stand for?"

"Creatures of Mystery," Melody responded happily. "We are the Creatures of Mystery crew, and we travel around in our Creature of Mystery carriage and look for, track, and study creatures of mystery."

The twins looked at the peculiar group of people they were sitting with and at the hand-painted rainbow-colored Volkswagen van full of strange equipment and decorations. This was a weird setting they found themselves in, and they had no idea what it had to do with botany. And they were curious about their local expert, who did not seem to be a part of this group. Even so, the twins were enjoying making new friends.

"There are actually C.O.M. Crews all over the world. We are just one of many," Harmony spoke up. "Right now, there are crews studying the Loch Ness monster, chupacabras, Irish elves, Area Fifty-one, alpine newt lizards, abominable snowmen, swamp creatures, and more."

"Yeah, like the crew studying hamsters," Chorus interjected, full of energy.

"Hamsters? Really?" Rip quipped. "Cory, hamsters are not creatures of mystery."

"Yes, they are!" the tall man huffed. "Haven't you ever seen a hamster chew something? How can they chew so fast? It's like lightning! That's mysterious, is it not?"

"No, Cory, it's not mysterious. It's—" Rip would have kept the silly argument going, but he was interrupted by the sound of an approaching vehicle.

The group of eight turned to see a khaki-colored Jeep drive up and park right beside the C.O.M. Carriage. The single occupant of the jeep turned the vehicle off; stepped out into the sand, approached the group slowly, and stood there staring at them for a few seconds with arms folded and one raised eyebrow.

"I suppose you are the C.O.M. Crew?" the jeep's driver finally stated.

"Yes, we are," Melody replied cheerfully. "And you must be Ranger Brock!"

The newcomer nodded. "Yes, Park Ranger Brock Hoverbreck,

to be exact."

His raised eyebrow went down, but his arms remained folded. He didn't look excited to be here. He was a very solidly built man with the stoic, chiseled face of an outdoorsman. He wore a crisp park ranger uniform, complete with a straight brimmed hat that matched the color of the jeep he had driven.

"Are those beanbag chairs?" The ranger asked, looking at the multicolored blobs the group was sitting on.

Melody silently answered, "Yes," with a smile and a nod, her cheerfulness completely intact.

"Yes, sir," Cory answered as well. "And the color they become depends on what kind of bean is put inside them."

"That's not right at all, Cory." Rip scowled.

"Yes, it is!" Cory retorted.

"No, it isn't! It's—"

"Never mind that," Ranger Hoverbreck interjected authoritatively. "Let me tell you right now how this is going to work. I am here to host you and serve as your guide in the Redwood National Forest, but you will follow my orders. If I tell you to stop, you stop. If I tell you not to touch, do not touch. If I tell you to stand in place and do one hundred jumping jacks or flap your arms like wings and hoot like an owl, you jump the jacks and you give the hoot. Do you understand?"

The group nodded.

"I love nature." Hoverbreck continued. "But I don't care for people who tramp into the serene forest making lots of noise, throwing away trash, or destroying the fauna. I am not hosting you because I want to; I'm hosting you as a favor to my favorite nephew. He read an article about a C.O.M. Crew online, and he has become a fan of yours. He loves all the research you do on creatures of mystery around the world. He enjoys the stories you tell about your adventures seeking them out. So, when he found out you wanted to send a crew to the redwoods of Northern California to look for Bigfoot, he begged me to volunteer to be your guide. That is why I am here—the only reason I am here."

Ranger Brock stopped and let out a deep breath. "With that out of the way, are we going to have any issues on this trip?"

Melody answered for the group, "No, sir." She smiled. "We will happily follow your orders, and we are very grateful that you came out to host us and serve as our guide."

Brock gave a curt nod, and the C.O.M. Crew began putting the beanbag chairs back in the C.O.M. Carriage. Once done, they pulled all the equipment they would need to go squatchin' and camp out. As they did, Ranger Hoverbreck walked over and gave Blaine and Tracey a good once-over.

"Are the two of you with this C.O.M. Crew?" he asked.

"No, sir," Blaine answered. "We're here to study some of the plants in the area."

Upon hearing this, the ranger's face relaxed.

"Well, that's good to know," he responded. "At least some people around here aren't crazy."

Ranger Brock paused as he looked over to where the beach ended and the forest began. "Have you learned about the spruces yet?" he asked.

The twins shook their heads.

"Come with me." Hoverbreck led the twins closer to the forest and pointed out a specific tree. "That tree is a Sitka spruce. They are found from the coast of Northern California all the way up into Alaska. Actually, they are the official state tree of Alaska. Here in California, they grow on the sandy coasts and rocky cliffs of the beaches of the Great Northwest. All of the scrub plants on the beach and these salt-tolerant spruces that grow near it act as barriers for the plants that live further inland, such as the redwood, which would be damaged by the saltwater spray of the ocean.

"Sitka spruces have shallow root systems because of where they grow. So, if a strong winter cyclone hits the coast, they can be blown down. The Sitka is the largest spruce tree in the world and the fifth largest conifer in the world. They can grow to be over three hundred feet tall and fifteen feet wide. This is mainly because of the moist ocean air and summer fog, which provides the tree with plenty of water to grow."

The Sassafrases were growing more comfortable with Ranger Brock Hoverbreck. He was authoritative and a bit intimidating, but like all the rest of their local experts, he had a good heart. They watched him come alive as he was sharing about the subject of his expertise.

"The thin, grey bark of the Sitka spruce can flake off in small circular plates." The park ranger continued. "The crown of the tree is shaped like a cone, and many of the older trees do not have branches near the ground. The needles are stiff, short, and specially designed for the conditions that the tree grows in. The fog condenses on these needles, and the tree is able to absorb the water from it. The tree's needles can also absorb the calcium and phosphate it needs from the ocean spray.

"An easy way to identify these trees is to grab a branch. Since the tough pointy needles stick out in all directions, it will hurt when you grab the Sitka's branch. The Sitka spruce also has cones, which are several inches long and cylindrical. The cones house their seeds. These tiny, black seeds have brown wings that help the wind carry them to a new place where they are able to sprout and grow into a new tree."

As Ranger Hoverbreck finished giving information about the Sitka spruce, a bald eagle shot out and took flight from the very tree at which the three had been looking. The twins missed taking a picture of it, but they managed to get a nice shot of the Sitka spruce.

"Beautiful tree. Beautiful bird. Beautiful forest." The big ranger sighed and smiled slightly. Then, just as soon as it had appeared, the smile disappeared as the ranger turned from the forest to face the ragtag group of hippies.

"Is everybody ready?" Hoverbreck asked.

"Yes, sir, we are!" Melody Albermully answered merrily.

Cones of a Different Origin

The Sassafras twins found themselves hiking through the most beautiful forest they had ever been in. That was impressive, considering the jungles and forests they had already experienced this summer. They were in the middle of the pack, and though they weren't official members of the C.O.M. Crew, Sam had given them camping gear to carry.

Melody, Sam, Rip, and Ranger Brock were in front of them, with the ranger leading the way—behind them were Harmony and Cory. Ned the driver had stayed back at the C.O.M. Carriage. The twins found out that he rarely joined the crew on their expeditions. He was going to park the rainbow-colored Volkswagen by the roadside to see if he couldn't make a little extra money by selling seashell jewelry.

As they walked, a beautiful blanket of dead brownish-orange pine needles covered the gently winding trail. It curved around trees and rose and fell with the slightly hilly landscape. There was a wide assortment of ferns, shrubs, flowers, and trees, but by far, the most breathtaking of all were the forest's namesake—the towering redwood trees.

CHAPTER 14: CALIFORNIA CREATURES OF MYSTERY

They rose high into the sky like silent giants watching over a magical timberland. The group of eight walked at a decent pace deeper and deeper into the forest. The twins could hear Melody and Ranger Hoverbreck talking about a specific spot that she had in mind to camp—a spot that was near several recent Bigfoot sightings.

"Yes, I know the spot you're talking about, and it is a fine place to camp," the ranger was saying. "But, as for all the Bigfoot sightings, I do not consider any of those factual. I interviewed most of the people who reported those sightings, and I would classify them as less than reliable. In my expert opinion, you are chasing the wind, Miss Albermully."

"Pluh," Cory grunted in disgust from behind the twins. "His expert opinion," he said sarcastically but only loud enough for only those close by to overhear. "Doesn't he know who the real experts are here? Doesn't he know that not only does Bigfoot live in this forest but so do trolls, ewoks, fairies, and extra-terrestrials?"

Harmony heard her brother, as did the twins, and she let out a good-natured laugh. The twins didn't know how to react. They still had no idea what to think about this whole creatures of mystery thing.

The group hiked on for another hour or so, before finding the camping spot Melody had been talking about. It was a fairly wide-open and flat spot with a small stream running through it. A glimpse of the blue sky could be caught here, which was rare because of the towering tree cover. The spot was barricaded off on one side by several fallen redwoods. The ancient trunks lay stacked into a picturesque immovable wall. All other sides of the spot were thick with trees of different sizes, and trails seemed to lead off in every direction.

It was beautiful, but a little creepy. It was quickly decided that this was where they would camp. Sam, whom Melody had introduced as the tech specialist, was also obviously the camping expert of the group. He gave directions for how and where to set up all the camping equipment, and when he did, everyone jumped to it.

Blaine and Tracey were responsible for putting up tents alongside the Albermully sisters, which was fine with them, since Melody and Harmony seemed to be the most pleasant of the group.

"So, how did the two of you get into all this in the first place?" Tracey asked the sisters.

"Michael LaGrange," Harmony blurted out dreamily.

"Well, yes, there is Michael LaGrange." Melody nodded. "But it all actually started because of Tiffany Cratcher."

"Tiffany Cratcher?" Tracey asked.

"Yep, she was the unelected leader of a group of high school girls named the Soccer Chicks. They were girls from the soccer team who were notoriously popular and infamously mean. They were especially unkind to new students, who included Harmony and me. Our parents were traveling musicians and artists, so we moved around a lot and never stayed in one place for very long. The very first day of school, at lunchtime in the cafeteria, Tiffany, followed by her entourage of Soccer Chicks, walked up and knocked our lunch trays out of our hands. Then, she sarcastically said, 'Oh, no, looks like you've spilled your lunch. That's too bad!'"

Tracey shook her head. "That's terrible."

"It was terrible," Harmony agreed. "We stood there with our lunch plastered all over the front of us, listening to everyone in the cafeteria laughing at us. That's when Michael LaGrange showed up." As Harmony finished, she just stared off into the forest with a wide smile and googly eyes.

The twins looked at each other as they thought, "Another case of the googly eyes!" Googly eyes were what the twins' father had always called the look on a person's face who was smitten with someone. The googly eyes had happened already this summer to Blaine and Tracey, so they understood."

Melody chuckled at her sister before continuing the story. "Yes, that's when Michael LaGrange showed up. He knelt down and helped us pick up the salvageable parts of our lunches. Then, he stood up and called out to Tiffany Cratcher in front of everyone. 'Not cool, Tiffany. You and the Soccer Chicks think you rule this school, but you don't! This is no way to treat new students or people in general. You're a joke, Tiffany. Everyone should be laughing at you instead of at these two girls!'

"There was a momentary silence after Michael said this, and then everyone switched from laughing at us to laughing at Tiffany and the Soccer Chicks. This was a crushing blow to Tiffany because she had never been laughed at before. Plus, she had a huge crush on Michael LaGrange. He was a smart, good-looking guy, and he just happened to be the captain of the boys' soccer team."

"But how does all this tie into you two becoming part of the C.O.M. Crew?" Blaine asked.

"Well, after that day in the cafeteria, we Albermullys became good friends with Michael," Melody responded. "It turned out that even though he was the ever-so-trendy captain of the soccer team, he was also into history and art, like us. He was especially interested in the history of mythical creatures. The three of us started a C.O.M. Club at the high school, which became wildly popular. We made a pact that when we finished high school and college, we would start a C.O.M. Company together."

Melody paused, and the smile slowly faded from her face.

"But sadly, the summer after our junior year. Michael disappeared while on a camping trip with his family. Search-and-rescue teams were sent out, but after weeks and weeks of looking, they never found him. Needless to say, it was hard for us to accept. I had always viewed Michael as a sort of older brother. So did Chorus, but it was much harder for Harmony because she'd fallen in love with Michael."

"And I'll never love again," Harmony interjected soulfully,

with watery eyes.

"So, after college, even though Michael wasn't with us anymore, we started a small C.O.M. Company," Melody went on. "And, just like our C.O.M. Club in high school, our C.O.M. Company became a smashing success. We have been doing this for over a decade now. And as Harmony told you earlier, we have C.O.M. Crews all over the world, seeking out all kinds of creatures of mystery."

The Sassafrases were still undecided about the existence of creatures of mystery, but they were not undecided on the Albermully sisters. They liked them and hoped for their success.

The four worked together and finished erecting enough tents for everyone in the crew. Blaine and Tracey saw that Ranger Hoverbreck had built a safe fire pit in the middle of the clearing. Sam had set up an open-air tarp-covered tech station. Cory and Rip worked on whatever their project was out on the perimeter of the camping spot. They were not done yet because they were still arguing.

"No, Cory! The sensors can't get their feelings hurt."

"Yes, they can, Rip!"

"No, they can't!"

"Yes, they can! They are sensors, right? So, they must be sensitive, and you just keep jamming them into the ground with no regard for their feelings!"

"Cory, they're not called sensors because they have emotions. They are called sensors because they can sense any heat or movement that approaches our campsite. It's—"

"That's enough!" Sam called out, interrupting the argument.

"There are plenty of sensors around the perimeter. Let's all gather around the tech station, and I will brief everyone about our new C.O.M. Cones." Rip jammed the last sensor into the ground with emphasis, scowling at Cory.

The sensors were metal staffs, which were around six feet tall and had small illuminated beacons mounted on their tops. By being placed around the perimeter, they could detect any approaching creatures of mystery. They also placed a kind of security fence around the C.O.M. Crew campsite.

The Sassafrases walked with the Albermully sisters over to the tech station as everyone, except Brock, congregated to hear what Sam had to say about something he had just called "C.O.M. Cones." He had one displayed on a camping table, and everyone who was a part of the C.O.M. Crew seemed to know exactly what it was and what it was used for. The twins, on the other hand, had never seen anything like it. They were all looking at a fairly large, cone shape made out of clear plastic. It had several different gadgets and wires attached to it.

"I know this model looks pretty much like the last C.O.M. Cone we used, but there are a couple of upgrades that I wanted to inform everyone about," Sam addressed the group.

The tech specialist picked up the cone and placed it over the top of his head and shoulders. It seemed to be pretty lightweight and looked comfortable to wear. It fit securely with small padded braces that rested on the man's crown and shoulder blades. Two black, plastic-coated wires with a button-covered controller at the end of them now hung down from the cone in front.

Sam secured the wires snugly to his arms via Velcro straps and grabbed a controller in each hand.

"Just like before," Sam's voice could now be heard over a speakerphone. "We have radar."

As the techie said this, he pushed a button on the controller. This caused a small rounded propeller to slowly spinning at the very tip of the cone, which the twins assumed was the radar.

"We still have our ever so helpful data screens," Sam said

as he pressed another button. This time, a screen in the side corner popped up with Bigfoot information on it. This was a screen that was much like the screen on Summer Beach's translucent tablet, where the images could be viewed from both sides of the see-through cones.

"We also still have floodlights." The next button Sam pushed turned on bright lights, even though it was still daylight. "And night vision capabilities because you all know how often we track at night."

Sam took a breath before saying, "Now for the upgrades."

The techie kept working the handheld controllers. "An interior light."

The group oohed and aahed as a greenish-yellow light clicked on, illuminating Sam's face on the inside of the cone.

"A pulse oximeter," Sam informed. "This will enable us to monitor our heart rates and oxygen levels, ensuring everyone's health. And, last but not least, per Rip's request, these new C.O.M. Cones are equipped with tranquilizer darts."

Rip pumped his fist in elation at this announcement. Sam touched another button. Small barrels on the bottom sides of the cone aimed. Another push of a button, and *swoop*! A dart shot out and stuck firmly into a nearby tree. The C.O.M. Crew along with the Sassafras twins oohed and aahed again.

Ranger Hoverbreck, from his position at the fire pit, looked less than impressed. He just shook his head, more like an exasperated babysitter than a thrilled host.

With Sam's demonstration complete, the group dispersed from the tech station. The ranger took Blaine and Tracey aside and asked, "I thought the two of you were here to learn about plants, not gaggle over all of this ridiculous gadgetry?"

The twins didn't know what to say.

"C.O.M. Cone?" Hoverbreck asked. "Really? The forest already has plenty of cones. Look around! Do you see all of the evergreen trees and shrubs with needles for leaves?"

The twins nodded as they looked around, seeing what the ranger was pointing out.

"These are called conifers or coniferous plants. The Redwood National Forest is a coniferous forest, which means it is full of these conifers. Conifers are gymnosperms, which means naked seeds. In other words, they don't have flowers that develop into fruit like angiosperms do; instead, they produce a dry fruit known as a cone. Are you following me?"

Blaine and Tracey nodded again.

Ranger Brock continued, full of gusto. "Conifers produce these cones to protect and disperse their seeds. Essentially, the cone is the reproductive structure of a coniferous plant. There are male and female cones, both of which have scales, but they differ in color and firmness. The male cone is typically soft, green, and small. In fact, many of us never even notice it, but it is very important. You see, the male cones produce the pollen for the conifer tree. Without that, the seed would not be able to develop.

"Under each of the scales of this cone is a sack that contains the yellow powder known as pollen. When the conditions are right, the male cone opens, and the wind disperses the fine particles. Some of the pollen reaches the female cone to fertilize it, which causes it to grow and develop seeds. The mature female cone is typically larger, harder, and browner in color. It's what we normally refer to as a pine cone, but even these cones start out soft and green.

NAME: Cone
DESCRIPTION: The reproductive structure of a coniferous plant.

"So, once the seed is fertilized,

the female cone darkens and hardens as the seed matures. This process can take a year or more to occur. As the seed matures inside the cone, a type of glue called resin helps the scales remain tightly closed. When the temperature and humidity are right, the cone opens, and the seeds are dispersed by the wind. There are many different shapes and sizes of cones, and each is specific to the type of tree from which it comes. Scientists use these cones as a means of identifying a conifer tree."

The more Ranger Hoverbreck talked about science, the more relaxed he became. Now that he was finished giving the twins information about cones, he was even smiling. The twins smiled and took a picture of a nearby hanging pine cone.

Later, after a nice meal around the crackling campfire, Melody laid out the plan of action for the group. "Though there have been several sightings of Bigfoot in the daytime—like the famous Patterson-Gimlin sighting. We will wait to seek this creature of mystery out at night. After my research, I believe that this is when Sasquatches are most active. So, we will wait for darkness to fall, which should happen in about an hour. Then, we will gear up and get 'squatchin'."

Melody's announcement gave the Sassafras twins a chill. It was one that was made up more of excitement than fear. Sasquatches, dark mysterious forest, tranquilizer dart-shooting C.O.M. Cones—this was bound to be terrorizingly fun.

Chapter 14: California Creatures of Mystery

Chapter 15: Sasquatch Sighting!

Exploring the Redwoods

The eerie lights of the C.O.M. Cones cast ominous green light and shadows across everyone's faces as the group slowly walked through the dark forest. Tracey felt more like she was exploring the moon than the woods. Her C.O.M. Cone fit fine and was working perfectly, but it made her feel like an astronaut in a space suit.

Blaine, however, thought they all looked more like weird giant lightning bugs than astronauts as they bounced through the darkness with the green-lit cones over their heads and shoulders and the lower halves of their bodies walking unlit and virtually invisible. Everyone in the group was wearing a cone except Brock Hoverbreck. The big park ranger had refused the offer from the C.O.M. Crew to use one as he folded his arms. His attitude toward the Bigfoot-seeking hippies was still skeptical.

The Sassafras Science Adventures

The group of adventurers had been quietly walking through the nighttime redwood forest for several hours now, with Rip the professional tracker, or hunter as he called himself, in the lead. There had been one instance when Rip had thought he had picked up the trail of a Sasquatch, but nothing had come of it. So, the hours of hunting had mainly consisted of quiet uneventful green-lit trudging.

Blaine used his C.O.M. Cone to check his pulse and oxygen level—pulse eighty-two, oxygen level ninety-eight percent. The levels were both good. Now he clicked his data screen on. It immediately illuminated inside the clear surface in front of him. Even though it was only inches away from his eyes, he could clearly see the words and images displayed on it. As he looked at the picture of the Sasquatch, he wondered again if these things were real or not. The twelve-year-old doubted they were.

Blaine sighed and clicked the data screen off again and gazed out into the forest using the cone's night-vision capabilities. His eyes swept from left to right and right to left. Wait, what was that?

He was fairly sure he had just seen a pair of eyes staring at him. He looked back a little to the left, where he thought he had spotted the eyes . . . they were gone. They were not there anymore. Was his imagination playing tricks on him?

There they were again—two big round wide eyes staring right at him. Blaine was about to alert the group when suddenly the eyes rose up several feet into the air, and a toothy gaping mouth just under the eyes let out a loud rumbling roar that shattered the silence of the night. The sound was so terrifying that Blaine literally fell to the ground.

"Bigfoot!" someone from the group shouted.

"No! Big bear!" Ranger Brock's voice strongly corrected. "Everyone get face-down on the ground now! Don't move!"

"No, Ranger!" Rip shouted. "Let me shoot it with a

tranquilizer dart!"

"Get down right now, Rip," Hoverbreck ordered in a stern voice that would not be argued with. "Don't you dare use those tranquilizer darts. I have bear mace, and I am deploying it right now!"

Tracey didn't know what the rest of the crew was doing, but she immediately obeyed the ranger's order. She shot to the ground face-first. Blaine, who was already on the ground, made sure he was as low as he could go.

"Kip-shhhhhhhwick!" went the sound of Ranger Hoverbreck pressing the button on the aerosol can. The brave ranger started shouting at the growling bear and banging his flashlight against a tin coffee cup he had with him. He was trying to make enough noise to get the bear to turn on his heels and go off in the other direction.

The Sassafrases lay there with their faces in their cones and their cones in the dirt. They were blind to what was happening with the huge bear. Was he approaching? Was he hovering over them even now, ready to swipe them with his long claws?

Ranger Brock shouted and made as much noise as possible while everyone else remained still and quiet. It continued on like this for several minutes, and then slowly but surely, the ranger's noises grew quieter until he finally stopped. There was a long second or two of complete silence, and then Hoverbreck told everybody it was okay to get back up.

The bear was gone. A collective sigh of relief was released, and without even needing to discuss it, the group knew that it was time to call it a night. There would be no more squatchin' tonight.

The next morning, after dreams of Sasquatches sipping cups of café au lait and bears driving sports cars, the Sassafras twins awoke early. "Zzzzzip," went the sound of their tent as they opened it up.

"Whoa, Tracey!" Blaine exclaimed groggily but excitedly.

"Look at all of the fog this morning!"

Blaine stepped out of the tent, allowing Tracey the chance to peek out. Tracey nodded—groggy, but excited as well. The fog was so thick the twins could barely make out the form of the next tent over, which they knew was only a few feet away. They had never seen anything like this.

They slowly made their way through the fog toward the fire pit, where they planned to start a fire to warm up a bit. When they reached the fire pit, they saw that it was already glowing orange with flame.

At first, they didn't see anybody sitting or standing next to the pit. They saw only the flame. As they looked around a little more in the fog, they caught a glimpse of a lone figure sitting still and quiet on a stump a few feet back from the pit. It was not a Sasquatch or a bear. It was Ranger Brock Hoverbreck, last night's hero.

"Morning, Blaine and Tracey," he greeted, neither high nor low. "What do you think about all the fog this morning?"

"It's amazing," Tracey answered. "Is this normal for the redwood forest?"

"Yes, ma'am, it is," the ranger answered. "Fog accounts for a fourth of the needed precipitation for the coastal redwoods in this forest."

The twins found rocks to sit on and joined the big ranger—one on his left and one on his right. The fog was thick on every side, but the twins found that if they looked up, they were able to see the tops of some of the towering redwood trees.

Hoverbreck saw the twins peering upward, so he joined them in their gaze with a respectful look in his eyes.

"Redwood trees are the giants of the forest," he stated. "They can grow up to twenty feet in diameter."

Blaine exclaimed, "That is big enough to drive a car through if it was tunneled and carved out!"

"Yes, it is," Ranger Brock replied. "They can grow up to three hundred feet tall."

"Three hundred feet high!" Tracey cried. "That's taller than the Statue of Liberty!"

Hoverbreck nodded. "The trees in this forest can live to be up to two thousand years old."

"Two thousand years old!" Blaine blurted out. "That means some of these trees were here when Ancient Romans walked the earth!"

"Right again," the ranger verified.

The Sassafras twins were truly amazed and awestruck at the magnificence of these trees.

"Yes, these trees are ancient, but because of pollution, logging, and habitat loss, they are also endangered," Ranger Brock said solemnly.

"There are two main types of redwood trees in the United States." The big man continued. "The giant sequoia, which is found in the Sierra Nevada mountains and the coastal redwood, which you are looking at right now. It is found here on the coasts of Northern California. The giant sequoia tends to grow wider, while the coastal redwood tends to grow taller. There is a third type of redwood tree, called the dawn redwood, but it grows only in certain valleys in Sichuan, China."

Tracey cast a little look her brother's direction. She had been to Sichuan, but Blaine had not.

"Redwoods are coniferous trees." Hoverbreck went on. "So they stay green all year round. They also produce cones to protect their seeds as they mature instead of flowers or fruit. Redwoods have a thick cinnamon-red bark that can be up to twelve inches thick. This makes the tree extremely resistant to fire damage and water-rot.

The trees also secrete a special chemical that bugs and decay-causing fungus don't like to eat. So, their wood sticks around for a very long time.

"Their leaves are small, scale-like, and spirally arranged around the branches. Because of their height, only the top half of the tree has branches. Redwoods have a shallow root system that extends quite a distance around the tree. The roots of one tree often intertwines with the other trees, which gives them more strength and stability."

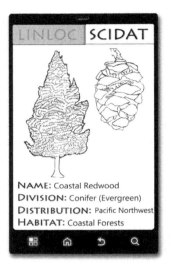

NAME: Coastal Redwood
DIVISION: Conifer (Evergreen)
DISTRIBUTION: Pacific Northwest
HABITAT: Coastal Forests

The ranger's face broke into a full smile. The man found great satisfaction in talking about his beloved forest, and the twins loved hearing about it. Then, like all the times before, when he was done giving information, the ranger's smile disappeared and was replaced with a look of exasperation. Evidently, he was not looking forward to another day with the C.O.M. Crew.

The twins looked up once more, and snapped a quick picture of the awe-inspiring coastal redwood trees.

The Sassafrases liked Ranger Hoverbreck and were glad that they were becoming friends. Brock shared some of his coffee with the twins, and the three sat quietly conversing around the crackling morning fire. They were soon joined by a flurry of sleepy-eyed, bed-headed, yet still energetic, pursuers of mystery.

"Get a load of this fog," Rip marveled to Sam as they walked up together. "This stuff is going to be hard to hunt in."

"Track in," Melody corrected the camo-wearing man with a smile as she and Harmony walked up to the fire behind Rip and Sam.

"Isn't the fog hauntingly beautiful, though," Harmony chimed in. "It floats in off the Pacific Ocean and blankets the ancient forest. It shrouds the sentinel-like redwoods in mystery and creates the perfect habitat for the enigmatic creature we call Bigfoot."

"It is beautiful," Melody agreed. "And I have a good feeling about today. I think we are going to get a good sighting of our creature of mystery!"

"Oh, you've had your good feelings before on days that we've seen nothing," the always combative Rip reminded her. "Good feelings do not mean anything. It's all about precise tracking."

"Ha!" Melody laughed. "You finally said tracking instead of hunting!"

Rip shuffled a bit. "Good tracking . . . is hunting."

"I'm not going to argue with you, Rip; that's Cory's job." Melody smiled. "By the way, where's Cory? Wasn't he sleeping in that three-man tent with you and Sam? Didn't he wake up with you guys?"

"Yeah, he was in our tent, but he got up way before Sam and I did. I thought he was out here somewhere by the fire. He isn't?"

"No, he's not," Melody answered, her smile fading a little.

"Well, it's really foggy. Maybe he's here, and we just can't see him."

"Rip," Harmony interjected. "He's not here. It's foggy, but it's not that foggy. We can see everyone here around the campfire. You, me, Melody, Sam, Ranger Brock, Blaine, and Tracey . . . but no Cory."

"Well, if he is not in the tent, and he is not by the campfire, where is he?" Rip voiced the obvious question.

A murky silence fell across the group as the next obvious question hit them all at once. Did Cory's absence have something to do with the bear from last night? If that was the case, their situation had gone from laid back to dire.

Ranger Hoverbreck wasn't the biggest fan of the C.O.M. Crew, but he was a professional. He did not want anything bad to happen to them or to see any of them go missing, especially in his forest. A look of concern mixed with sincere sympathy formed on his face as he stood up with a plan.

"We are going to find Cory," he announced. "Rip and I will search the perimeter of the campsite for any tracks that Cory may have made. When we find them, we will all form a line or human chain, standing close enough to each other in the fog that we can still see the person to each side of us. Rip and I will be on the ends. We will then walk carefully through the forest, looking for Cory and shouting his name until we find him. The fog is beginning to clear, so our visibility should continue to get better and better. Sam will wear a C.O.M. Cone to use its radar and other gadgets, but the rest of us will use our God-given senses to search. Is everyone on board with this plan?"

Everyone nodded

"Then let's get to it! Right now."

Sam retrieved a C.O.M. Cone from the tech station and secured it to his frame. Hoverbreck and Rip scoured the campsite's perimeter, looking for Cory's tracks. At the same time, the rest of the group gathered their resolve and began forming the human chain that the ranger had described.

Blaine and Tracey were still a bit shaken from last night's encounter. They hoped they didn't run into another mammal of that magnitude, but more than that, they hoped they could find Cory.

"I've got something!" Rip shouted out. "I've found tracks that are more than likely Cory's. And from what I can see, it looks like he walked off into the woods alone."

The group hurried through the redwoods, searching in a line. They started from where Rip had spotted the tracks. Sam was in the middle, wearing his cone. Blaine and Tracey were on his left and right. Melody was to Blaine's left, and Rip was left of her. To Tracey's right were Harmony and Ranger Brock.

Blaine looked to his right, and Tracey looked to her left. Each could see Sam, but the twins couldn't see each other because of the low clouds. The Pacific Coast fog was such a strange phenomenon.

"Cory! Cory!" the group shouted over and over.

It was a strange thing to be searching in such thick fog because at one moment, all that was visible was the grayish-white of the fog and the next moment the trunk of a massive redwood tree would be only inches away from your face. How would they find Cory in all this fog?

Tracey appreciated Ranger Hoverbreck's more natural approach, but she could not help but think how nice it would be to have one of the C.O.M. Cones on right now. She felt like she could not see a thing. At least with a cone she would have radar and floodlights.

She looked over at Sam in his C.O.M. Cone and wondered if he was picking anything up on his radar. It didn't look like it. He was just walking straight ahead with no dramatic movements or noticeable emotions.

Tracey swung her head over to the right to see how Harmony was doing. The blonde-headed Albermully sister just called out vigorously for her brother. Tracey hoped, especially for Harmony's sake, that they would find Cory alive.

"Cory! Cory! Can you hear us?" Tracey cried out as she looked to her left again.

"Cory! Cor—" Tracey's sentence stopped as abruptly as she did.

Sam . . . was gone. The C.O.M. Cone he'd been wearing was now lying on the forest floor with a big crack on one side.

Magic Vanishing Mushrooms

Blaine had heard something, but he hadn't seen anything. He had heard a slight rustling of fern leaves, a crack, and a soft grunt, but he hadn't seen the source of the sounds. He quickly walked over to his right to the place where he thought the sounds occurred, and what he saw immediately sent tremors through his heart. Lying on the ground was Sam's busted C.O.M. Cone with no Sam in it.

Suddenly, a figure burst through the fog toward the spot. Blaine jumped into a defensive position, but he didn't really need to because the figure was Tracey.

"What happened to Sam?" she asked. "Did you see anything?"

Blaine shook his head.

All of a sudden, the twins spotted it at the exact same time. Right there in the soft dirt, next to where the cracked C.O.M. Cone, was a footprint—a big footprint. A chill ran through the Sassafrases' veins that was full of fear with no excitement. There was a Bigfoot on the loose in these woods, and it had just abducted Sam.

The twins' fear compelled them to sit down, but they fought that compulsion and started shouting out the names of their friends. Ranger Hoverbreck, Rip, and the Albermully sisters all responded that they were on their way.

Just moments after they began calling out, Melody was approaching them, but to the twins' horror, before she could reach them, a huge nearly seven-foot-tall hairy ape-like creature darted out of the fog and swept the woman right off the ground. Then, as soon as the thing had appeared, it disappeared with Melody in its grasp.

The twins now fall to the ground and started shivering. They had just seen Bigfoot with their own eyes. Harmony, Rip, and the ranger all ran up to Sassafrases' spot and found them sitting down.

They were about to ask what was wrong when they simultaneously spotted the footprint and the cracked cone.

"A Sasquatch!" Rip exclaimed.

"Did it take Sam?" Harmony asked.

The twins nodded. "And Melody," they added gravely.

Harmony's entire countenance fell upon hearing this.

Ranger Hoverbreck knelt and examined the huge footprint.

"Well, it's not a bear," he finally replied.

"Of course it's not a bear!" Rip blurted. "It's a Sasquatch, and it's—"

Rip's sentence was cut off by a loud rumbling roar. It was a roar that was reminiscent of the bear's roar from last night but different. This roar was a little higher pitched and more human sounding. Not friendly or normal human-sounding, but mean and wild human-sounding. The sound caused the Sassafrases to jump to their feet, and before they really knew what was happening, everyone was running off into the forest in different directions.

The twins managed to stick together as they plunged through the foggy green like wide-eyed stampeding cattle. They crashed aimlessly and urgently through underbrush, around tree trunks, and over gnarly roots, with no idea which direction they were going or where they needed to go. They were only thinking that they needed to GO!

All at once, they found themselves running up on a fallen log. It was a huge redwood tree that had toppled over. It had landed over a small gully, and rested on the gully's opposite embankment, creating a kind of tree-trunk bridge.

The twins scampered up and over with no real worry of falling off because the redwood was big enough to create a fine bridge. Just as they were running directly over the deepest part of the gully, a single figure they recognized crossed their path. Rip was running up the dry gully underneath their log bridge, just as urgently as they were running over it.

Suddenly, stepping out of the fog at the edge of the embankment, there it was again—the Bigfoot. The creature was below the twins and above Rip. It turned its big hairy head and looked right at the Sassafrases up on the log. Their hearts screamed. Was it going to come for them?

Instead of going for them, it jumped down into the gully and tackled the fleeing camo-wearing C.O.M. Crew member. Before Rip even knew what hit him, he was carried off feet first into the foggy forest.

Blaine and Tracey sprinted for the end of their log bridge, but as they did, they realized this end, unlike the end they had entered on, was sticking up a considerable distance off the ground. This, however, didn't stop the twins. They weren't going to stay exposed up on the log bridge with that Sasquatch so close.

Without stopping, they jumped off the end of the fallen redwood down into a swath of fronds. They crashed down into the leaves and then immediately scrunched down and focused on staying quiet. This was harder than it seemed, since their hearts were beating like rock-and-roll drummers. Plus, their breathing was coming in unstoppable gasps and heaves.

The Sassafrases sat low beneath the leaves and tried to calm down. They did not want to give their position away. Tracey tried to take deep breaths while Blaine covered his mouth with his hands, trying to muffle the sound of his gasping.

The forest around them was quiet, and after several long minutes, their heart rates and breathing slowed, joining the forest's silence. Without saying a word to each other, the twins' minds raced along the same thought line, "What was up with that Sasquatch?"

They had heard stories about Bigfoot before, and it was never painted as an aggressive creature. On the contrary, the Bigfoot was known as a recluse who tried to avoid human contact. That was what led to it being such a creature of mystery. Why was this Bigfoot attacking their group and taking them out one by one? Could it be that this Sasquatch was actu—

The twins' thoughts were cut off abruptly as the sound of something approaching them reached their ears. Whatever it was, it was moving slowly, and other than the occasional leaf rustle or snapped twig, it moved silently as well. It got closer and closer until the twins were both convinced they were about to get stepped on or, worse, eaten.

Now the Sassafrases were not only trying to keep their breathing quiet, but they were also trying to hold their breath altogether. The thing took one more quiet step closer, now enabling the twins to look up through the leaves from their hiding spot to see what or who it was.

It was Harmony! It was obvious that she was trying to be stealthy as she moved slowly through the waist-high plants. Blaine and Tracey were about to stand up and reveal their hiding spot to their friend when, in an instant, she was swept away in a flurry of growls and hair. One moment Harmony was there, and the next she was gone—the next member of the group to be taken by the Bigfoot.

The twins could not believe it. Harmony had been right there above them, and she, like them, had been unaware that the Sasquatch had been close enough to nab her. Now she was gone, and the twins were sure they were next. The question now was should they run or stay hidden. Either way, after seeing the Bigfoot's uncanny skills, the Sassafrases couldn't envision any way they would avoid abduction.

Using only facial expressions and eye movement to communicate, Blaine and Tracey decided to stay put. Long quiet seconds began to tick away, and the forest sat eerily silent as the twins stayed crouched in their hiding spot. Questions raced through their minds at the same pace as their speeding heartbeats. Were they the only ones left? How had Bigfoot taken everyone so quickly? What had it done with them after it had taken them?

Suddenly, there was only room in their minds for one single question: who was approaching right now? The Sassafrases could hear something moving through the brush toward them once more.

Again they held their breath. This time they completely expected a big pair of hairy arms to reach down through the leaves and swipe them up. A big pair of arms did swipe them up, but instead of hairy, they were khaki.

It was Ranger Brock.

"Blaine and Tracey," the big man whispered. "We're going back to the camping spot. Follow me."

The Sassafras twins very willingly obeyed the ranger's command. They stayed as close behind him as possible as he began to lead them through the forest to the camping spot. The fog still rested heavily among the trees, but it did seem to be lifting slightly, providing a little better visibility and a hint of hope.

It was apparent that the park ranger knew exactly where he was going as he weaved his way through the enormous tree trunks, taking lefts and rights on a rolling wooded trail with no hesitation. His boots thudded quietly but quickly on the ground. He was followed in step by Blaine and then Tracey, and then . . . wait, were those footsteps Tracey heard behind her? Big footsteps?

The Sassafras girl was too scared to look over her shoulder to see what might be there. She ran forward and pulled the back of her brother's shirt.

"Blaine!" she squealed. "Look behind me! Is there anything behind me?"

Blaine stopped. Tracey ran into the back of him. They both summoned the courage to turn and stare the stalking Sasquatch in the eyes, but it was not there. The forest behind them was hauntingly empty.

"There's nothing behind us, Tracey," Blaine assured her. "C'mon, let' keep going."

The twins turned back to follow Ranger Hoverbreck, but they were shocked to see he was no longer there. In the few seconds that Blaine and Tracey had turned around, the ranger had disappeared into thin air. He had become the next victim of the Bigfoot.

The Sassafras twins stood frozen and alone on the foggy trail. They had tried running. They had tried hiding, but nothing was working. If the Sasquatch had abducted the big and confident park ranger, it could take anybody.

The twins stood there trembling on the trail with their feet seemingly glued to the ground. Towering coastal redwoods rose skyward all around them, but even these stoic giants could not protect Blaine and Tracey from the inevitable.

A huge hairy figure emerged from the fog in front of them. It was Bigfoot. The mysterious creature stared the two twelve-year-olds down for a second and then began running toward them.

The twins could not run. They could not scream. They could only stand, hopelessly frozen stiff.

The creature ran with absolute silence and reached the children in a flash. It lunged with outstretched arms toward them, but before it made contact, another figure rocketed out from behind a redwood trunk and tackled the Sasquatch to the ground in a tumbling ball of hair, grunts, and pine needles.

It was Ranger Hoverbreck! The Bigfoot hadn't gotten him. Instead, he had gotten Bigfoot! The creature was much bigger than him, but Ranger Brock had no trouble at all wrestling the Sasquatch to the ground and then holding him there.

Ranger Hoverbreck started speaking to the ape-like creature as if it could understand him. "So, big fella, are you ready to be finished terrorizing this crazy C.O.M. Crew?"

Bigfoot growled.

"Well, ready or not, you *are* finished. You are not only finished with your terrorizing; you are also finished with your fooling. I know exactly who you are. You're not a Sasquatch at all. You can fool your own crew, but you can't fool me. Isn't that right . . . Cory Albermully!"

The twins gasped. Could it be? Could this terrible creature just be Cory dressed up in a Bigfoot suit?

Ranger Brock began grasping at the neck of the Sasquatch, and sure enough, he found the seam of the mask. The ranger pulled hard at the mask and ripped it off of the perpetrator's head. Indeed, this was a person in a Bigfoot costume, but it wasn't Cory Albermully.

It was Ned. Ned the van-driving, groovy-glasses-wearing, long-haired hippie of the C.O.M. Crew.

"Ned?" the ranger exclaimed in disbelief.

The twins couldn't believe it, either. Tracey's face wrinkled up. "But I thought you were selling sea shells down by the seashore?" she voiced.

Ned smiled like he'd been found out but didn't really mind. "No, little tike, I just said that so I could pull my prank."

"Your prank?" Blaine asked. "You mean dressing up in a Bigfoot costume and kidnapping us one by one?"

"Yep." Ned nodded. "That's right. It was pretty funny, now, wasn't it?"

"Funny?" Ranger Hoverbreck asked like the driver was crazy. "Have you been downing magic mushrooms? What you did was dangerous, and you scared your poor friends half to death."

"Magic mushrooms," Ned questioned. "No, no, no. I didn't have any mushrooms. I did mean to scare everybody, but I did not want it to be dangerous. I can show you where everyone is. I have them all tied and gagged over yonder in a hollowed out tree."

"Tied and gagged!" Hoverbreck exclaimed. "Ned, that's bad. That's really bad. Show me where they are now, so we can free them."

Ned hung his head like he was ashamed. He truly didn't mean for the prank to be dangerous. He stood to his big feet in his oversized Sasquatch suit and led the three through the forest to where the others were being held captive.

"What are magic mushrooms?" Blaine asked the ranger as they walked the trail behind Ned.

"As you may know, many mushrooms are poisonous and toxic. So you should never eat a mushroom in the woods that you don't know about," Ranger Brock replied with an edge of concern in his voice.

"Being poisonous prevents them from being eaten by animals. The fly agaric mushroom, for example, is toxic, and it is sometimes referred to as the 'magic mushroom' because its toxin acts as a psychotropic. Its poison is water soluble, so some people boil it for a long time and then eat it. It has a red cap with white spots so it is very easy to spot. The cap begins with an umbrella shape, but as it matures, the cap flattens out into a level plane."

The Sassafrases looked around as Hoverbreck was talking and tried to spot some mushrooms, but they didn't see any. Blaine mentioned this to the ranger, to which he responded, "The fly agaric fruits in early winter, so you won't be able to find any of those now."

The twins nodded in understanding. Later, when it was called for, they would just search through the archive app on their phones to find a good picture of this particular mushroom.

"The Redwood National Forest has all kinds of mushrooms." Ranger Brock continued. "It's a perfect habitat for these types of fungi because they prefer the moist woodland areas under the shade of trees. What we call a mushroom is really just the fruiting body of the fungus. In other words, what you see on the surface is only a small portion of the actual fungus. Underground, there is a network of cells called mycelium that form long strings, almost like roots. The mycelium can live for hundreds of years and send up multiple mushrooms each year.

"There are several different types of mushrooms, but the one everyone is most familiar with has the umbrella, or toadstool, shape. This mushroom has a stalk-like stipe with a cap on top. On the underside of that cap are gills that house the spores for the fungus. When the spores are mature, the gills release them, and the wind carries them to surrounding ground where they can form new mycelium and begin the cycle again."

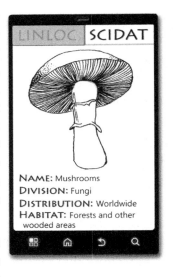

As they always did, the Sassafrases took precise mental notes on the topic of study at hand, so that they could regurgitate it into SCIDAT later.

After a short hike, the four reached the hollowed-out tree. A look of relief formed on the faces of Harmony, Rip, Melody, Sam, and Cory as they saw that the Sasquatch that had abducted them was actually just Ned in an oversized costume. After they had been untied and ungagged, their relief quickly turned to annoyance. They made it painfully clear to the driver that his little prank had sabotaged any chance they had at actually tracking a real Bigfoot.

Ranger Hoverbreck, however, looked amused. Watching the mystery hunters squabble among themselves was evidently entertaining to him. What the twins thought though is that this adventure that the gruff ranger had shared with the C.O.M. Crew had actually endeared them to him. And that, if he had to be honest, he would admit that he really liked them.

The group as a whole hiked back to the campsite, packed it up, and began the slow, quiet journey back to the rainbow-colored C.O.M. Carriage. Unfortunately, what none of the nine noticed before they left their campsite was the pair of big yellow eyes that had been staring at them from a hidden spot or the fresh footprints left by the fire pit—the really big footprints.

284

Chapter 16: The Glitch Is Gone

Joshua Tree National Park

It was ringing. It hadn't rung all summer, but it was ringing now in California. The Sassafras twins had just sent in their SCIDAT data and pictures, and almost immediately, Blaine's phone rang.

The twelve-year-old Sassafras boy answered. He assumed it was his Uncle Cecil, and he was right.

"Isn't that great news!" the eccentric red-headed scientist exclaimed loudly on the other end of the line. "Isn't it just radtastic! Splenderiffic! Superawesomeful! Isn't it just the greatest news, Train? Well, isn't it?"

"What are you talking about, Uncle Cecil? What news?"

"Well, what I just told you, of course!"

"You didn't tell me anything."

"I didn't?"

"No, you just said something was great, radtastic, splenderiffic, and superawesomeful, but you didn't tell me what that something was."

There was a pause on Cecil's end. "Oh, I guess I just thought it, but didn't actually say it. Well, okie dokie then. Goodbye, Train."

"Wait, wait! Uncle Cecil, aren't you going to tell me what was so great?"

Another pause ensued. "Oh . . . yes, yes, my boy. The glitch—it's gone! The glitch is fixed!"

"The glitch is fixed," Blaine repeated, excitedly.

"Well, almost," Cecil clarified. "It will be completely fixed after your next LINLOC location is reached and the SCIDAT data from that location is sent to the home computer here."

"It will?" Blaine asked him.

"Yes, it will," Cecil replied. "But there is one thing you and your sister will have to do differently."

"What will be different?" Blaine inquired hesitantly.

"You must travel to your next location without using the invisible zip lines."

"What?" Blaine gasped.

"Yessirree—President Lincoln figured out that this is the only way to put this glitch behind us. Check your LINLOC app, get the next longitude and latitude coordinates, and then travel to that location by any mode of transportation other than invisible zip line. When you get there, gather your SCIDAT data as usual and send it in. According to President Lincoln, once the SCIDAT data is in, the glitch will be a thing of the past!"

"That is great news, Uncle Cecil," Blaine responded cheerfully.

THE SASSAFRAS SCIENCE ADVENTURES

"Yes, it is, Train, yes, it is. It is simply flabbergasting! Explain all this to your sister, and I will see you two back in the basement before you know it!"

As soon as Blaine finished the call with his uncle, he told Tracey everything the scientist had shared. The twins were happy that soon there would be no more glitch in the invisible zip-line system. But before either opened up LINLOC to see what the next coordinates were a curious look formed on Tracey's face.

"Does it really matter?" she asked her brother.

"Does what matter?"

"If the glitch is fixed or not," Tracey answered.

"Of course it matters," Blaine retorted. "We don't want any glitches in the system, do we?"

"No, I guess not." Tracey sighed. "But all the glitch really did was force us to study science."

"What do you mean?"

"I mean, the glitch forced us to get all our SCIDAT data exactly right before we could progress through all our LINLOC locations and before we could zip back to Uncle Cecil's basement. Now that the glitch will be gone, it means we can make lots of mistakes with the information, and we can zip back home whenever we want. I mean, I hated science just as much as you did when we started this adventure, but now—I like it. I want to study it. I want to get all the SCIDAT info correct. I want to go to all the locations and learn about what is there. Don't you?"

Blaine nodded as both twins stood silently for a moment. Maybe the glitch really didn't matter.

Tracey opened up the LINLOC app. The next location was the Mojave Desert in Southern California—longitude 116° 3' 46.8" W, latitude 33° 58' 23.6" N. Their local expert's name was Symphony Douglas, and the next four scientific topics were Joshua trees, barrel cacti, creosote bushes, and paddle cacti.

Both Sassafrases instinctively pulled out their harnesses, helmets, and carabiners in order to zip, but then stopped short, realizing these things were not needed this time around. How exactly would they get to the Mojave Desert?

Just then, a big husky figure approached the twelve-year-olds. It was Ranger Brock Hoverbreck.

"Well, Blaine and Tracey Sassafras, it was nice meeting you. I actually enjoyed sharing this adventure with you and these silly mystery hunter friends of yours." The big man smiled, which was rare for him. "I'm about to get into my jeep and head back to the park ranger's station, but before I do, I want to gather everybody up to say a few words."

The twins nodded. They followed Ranger Brock over to the wildly painted Volkswagen van where all the members of the C.O.M. Crew were already standing.

"I think you are all crazy," Hoverbreck said roughly. "But with just a few exceptions, you all followed my lead and my rules in the park just fine. I am sorry that we didn't find you a Sasquatch—at least not a real one."

Everyone looked at Ned, who shamefully hung his long-haired head.

"But we did find a bear. We did get to drink in the grandeur of the redwoods, and yes, we did bond with each other."

Everyone looked shocked that Ranger Hoverbreck had just said the word "bond."

"And because of this, I have an exciting opportunity for all of you. I have a cousin who is a park ranger down in Southern California at Joshua Tree National Park in the Mojave Desert. I gave her a call, and she said she is willing to host your group in her park if you are interested in doing research on another creature of mystery."

The C.O.M. Crew and the twins gasped with happy excitement.

"Evidently, there is a sort of creature that roams the hot sands of the Mojave, known as a Cactus Head," Hoverbreck said. "I don't really know much about it, and I'm not going to stand here and act like I am interested in it, but I thought you all might be. Should I call my cousin back and tell her to expect you?"

With bright eyes and no hesitation, the six members of the C.O.M. Crew answered, "Yes."

The next thing Blaine and Tracey knew, they were sitting on beanbag chairs, cruising down the highway in the cramped yet comfortable C.O.M. Carriage. The Albermully sisters had insisted the twins join them in their search for a Cactus Head, and the Sassafrases couldn't refuse. Especially since the Cactus Head creature had been spotted in the Mojave Desert, which was their next LINLOC location. Plus, by riding the C.O.M. Carriage they were following their uncle's instructions on how to get rid of the glitch.

So, off the Sassafrases went, down the California highway, with the ragtag group of hippies.

He had heard the scary stories. That's why he had invited his two buddies to camp out with him here in the Mojave Desert. His intention had been to scare them, but right now, he was the one who was scared.

Earlier in the day, they had driven in and set up their tents at a very remote campsite way out beyond any mapped roads. They had had a good time doing some hiking and rock climbing before the sun went down. They ate a good meal and shared several shivering laughs as they recounted the many tall tales of the Cactus Head around a glowing campfire.

The three of them had retired to their respective tents for the night. His two buddies had quickly fallen asleep. When the embers of the fire had dimmed down to dark gray ash, he had quietly snuck out of his tent and walked into the pitch-black desert a hundred paces from the campsite. He had hidden some supplies that would enable him to pull off an epic prank on his two pals.

He bent down and had begun to move the small pile of rocks the supplies were under when he was suddenly overwhelmed with the feeling that someone, or something, was watching him. Then, to authenticate his feeling of dread, the sound of deep heavy breathing had reached his tingling ears. Soon after that, a pungent smell reached his nose.

So, here he was, literally shaking in his boots. Was he imagining things? Had he heard one too many scary stories? Maybe that was just a strange passing desert smell. It had to be the wind and not heavy breathing he was hearing, right?

He strained his ears through the desert dark and listened as intently as humanly possible . . . and heard . . . nothing but silence. He breathed a deep sigh of relief. His mind was just playing tricks on him.

He moved the small pile of rocks and grabbed a red-blinking flashlight, for his prank. Next, he scooped up a bullhorn and laughed as he did, thinking about how much fun it was going to be to scare his friends with this. He then grabbed a strand of firecrackers, the last hidden item, and stood up from his crouched position.

He clicked the flashlight on and started walking back toward the campsite. The blinking light gave the desert around him a red glow in one-second intervals. His heart beat fast in excited anticipation, as he trudged back toward the tents. Once again, he heard labored breathing, but he dismissed his mounting fear quickly, thinking the sounds were coming from his own lungs. No, wait—he was breathing fine. The labored and heavy breathing was coming from . . . someone else.

Shock shot down his spine. It was not the wind making the sound. He stopped and turned his head slowly to the left, in the direction that the sounds were coming from. And there, tinted in red light, he saw . . . a cactus.

The sight of a cactus in the desert wouldn't have normally been a scary sight, but with all the stories and legends of the Cactus Head creature fresh in his mind, it was scary—very scary. He couldn't see the cactus clearly, though, because the light he held kept blinking on and off.

"Dude!" he screeched, slapping the flashlight, trying to switch it off blinking mode. "Give me a normal beam of light, dude!"

The flashlight did not obey. It kept blinking red. And, to his horror, every time it blinked, the cactus was one step closer to him. The cactus was moving. The cactus was breathing. The cactus had . . . a face. A face that was full of terrorized anguish. He wanted to run, but his feet seemed to be stuck to the ground.

The cactus got closer and closer. It was much taller than he was, and its head had all kinds of protruding prickly spines. It bore down upon him, bathed in red-blinking light, and now just mere inches away, the mouth on the face of the cactus opened up wide and let out a bloodcurdling scream that pierced the night. The scream was followed by a shout. A shout that sounded like the cactus was saying, "Criers!"

The shout of the cactus snapped him out of his frozenness, and he turned and sprinted toward his campsite. The hairs standing up on the back of his neck communicated to him that the cactus was chasing him. He dropped the firecrackers, and the bullhorn, and the blinking red light as he ran, and he began shouting at the top of his lungs trying to awaken his friends.

"Hey man, what's going on, man?" one of his buddies asked groggily as he pulled them out of their tents.

"This isn't very sick," the other said half asleep as he half-pulled, half-dragged them to the car.

Screams from the pursuing cactus could still be heard as he fumbled around with the car keys trying to get them into the ignition. Finally, he got the correct key in the starter slot. He cranked the car to life. He switched on the headlights. There standing directly in front of them now, illuminated by the vehicle's beams, was the Cactus Head.

"Man! It's the Cactus Head!" His groggy buddy was now fully awake.

"Not sick! Not sick! Not sick!" his other friend repeated.

He pressed down hard on the gas, the tires spun for a second in the sand, and then off they went. The Cactus Head reached out for the car as they sped by, but he swerved to elude the creature. The three heard one last anguished and eerie shriek before the horsepower of the engine provided a successful getaway. They left the Cactus Head behind them cloaked in the darkness of night.

Ned, who had been forgiven for his frightful folly, drove the Volkswagen van through the night, enabling the C.O.M. Crew to get some sleep and arrive at Joshua Tree National Park in the Mojave Desert early in the morning. As they pulled up to the park's visitor center, they saw a group of young elementary school students standing in a huddle around a female park ranger as she gave a monologue about the park.

The C.O.M. Crew piled out of the van and sauntered over to join the group of students. Before the twins heard the park ranger say anything, they noticed that she was standing by a large wooden sign that said, "Joshua Tree National Park." Next to the sign was a strange yet beautiful-looking tree.

"So, the park is actually named after the tree," the ranger was telling the children with loads of cheer and joy in her voice. "The Joshua Tree is a monocot, and is a part of the agave family, which is in the same group of flowering plants as grasses and orchids. It is native to the deserts of the southwestern United States. It has a tough woody stem and long evergreen tapered leaves. These leaves grow in a spiral fan shape at the tip of the stem that you see here." The ranger paused to point at the unique-looking tree. "The Joshua tree blooms in the spring if conditions are right, meaning that there has been the right amount of rainfall and at least one hard freeze. Freezes can cause the growth tip of the branch to be damaged, but research here in the park shows that this also stimulates the growth of a new branch along with a bunch of creamy white flowers. The Joshua tree is able to sprout from roots or branches, which means that after a fire it can reproduce quickly even without a seed.

"For pollination, the Joshua tree uses the yucca moth, which collects pollen as she lays her eggs in the flower. When the larva hatch, they feed on one or two of the available seeds. This is an excellent example of a symbiotic relationship between a plant and an animal."

The park ranger paused with a big smile and clapped her

hands once, before saying, "Well, kids, I know I just used a lot of big words, so do any of you have any questions about the Joshua tree?"

One little girl raised her hand and then asked, "What's your favorite color?"

Still smiling, the ranger raised an eyebrow and then responded, "Orange, but that wasn't really a question abou—"

"Are you married?" another girl asked.

"I . . . uh . . ." the ranger stammered.

"I have a rock in my shoe," a freckle-faced boy announced.

"Unicorns shoot rainbows out of their horns," the next child shouted.

Then the dominos began to fall.

"I sometimes eat dirt."

"I have to go to the bathroom . . . oh . . . never mind."

"My dad is stronger than a bear and a grasshopper."

"Can you make a paper airplane for me?"

"My shoe is untied."

"Are we there yet?"

"Look, I'm a ballerina!"

"Roller skates, bumble bees, volcanos . . . sugar!"

In an effort to rescue the park ranger from the children and steer the conversation back toward science, Tracey raised her hand in the back.

"Can you tell us how old this Joshua tree is?" the Sassafras girl asked.

The ranger smiled and answered Tracey's question, effectively bringing an end to the whimsical slap-dash comments of the young children.

"It is difficult to tell exactly how old a Joshua tree is because the trunk doesn't form rings like other trees do, but you can divide its height by the average yearly growth, which is about a half an inch, to get a rough idea of the age. Now, I will share with you a few of my favorite fun facts about the history of this wonderful tree," the ranger said with a smile.

"In olden days, American Indians used the dried-out leaves of the Joshua tree to make baskets and mats. They also used the flowers and roasted the seeds for food. During the settling of the West, the limbs of the tree were used as fence posts for corrals and fuel for fires. Nowadays, we refer to it as the 'Seussian Tree' because it looks so much like a tree right out of a Dr. Seuss book."

The twins chuckled at that last fact about the Joshua tree as they each took pictures of it and the teacher of the elementary class had all the children say "thank you" to the park ranger. The teacher then led the students off to another location, effectively leaving the ranger alone with the C.O.M. Crew and the Sassafras twins.

Evidence of the Legendary Cactus Head

"Good morning!" she greeted them all. "My name is Symphony Douglas, and I am a ranger here at Joshua Tree National Park. What can I do for you?"

"Do unicorns really shoot rainbows out of their horns?" Cory asked.

"Cory," Rip growled, elbowing his buddy in the ribs. "That's not why we're here!"

"But do they?"

"It doesn't matter right now! We're here about the Cactus Head, not unicorns."

"Oh, yes! Of course, you are the Creatures of Mystery Crew!" Symphony said cheerfully. "Cousin Brock gave me a call and told me you guys would be coming. I am so excited to have the chance to host you guys here at Joshua Tree!"

"We are really excited to be here," Melody replied on behalf of the group, reciprocating Symphony's cheer. "Your cousin, Ranger Hoverbreck, was a very good host to us up in the Redwood National Forest while we were looking for a Bigfoot, which, sadly, we were unsuccessful in finding. He mentioned to us another creature of mystery, the Cactus Head. We've heard of this creature before and know it is said to live somewhere in the Mojave Desert, but our knowledge is very limited. We'd love to hear all you know about it."

"Yes! Of course! I will be more than happy to help you with this! Oh, this is all so exciting!"

Symphony led her eight new guests over to the visitor center and then into a back room where she opened up a file cabinet and began pulling out maps, documents, reports, photographs, and more. Symphony Douglas looked a little like her cousin, mainly because she was wearing a similar-looking ranger's uniform, but she acted nothing like him. As a matter of fact, they seemed almost polar opposite in personality. He was skeptical and reserved. She was cheerful and excited—especially about sharing the Cactus Head information with the C.O.M. Crew. She seemed to believe in its existence and was overjoyed at the prospect of a crew researching it.

"The legend of the Cactus Head is a fairly new," the female park ranger said. "Unlike many creatures of mystery that have had reported sightings for centuries, the first sighting of the Cactus Head was only twenty years or so ago. There have been frequent reports and sightings of the creature, with at least one every other year. Most of the sightings have consistent facts associated with them. They all say the creature looks to be half cactus and half human. They say it moves at night and is often heard moaning or shouting out something like the word "criers." On top of that, all sightings have happened within the boundaries of the Mojave Desert."

The C.O.M. Crew listened raptly to every word Symphony said as they began pouring over the documentation she had laid out for them. Melody and Harmony focused on the maps that pinpointed where all the Cactus Head sightings had happened. Sam and Ned looked through the photographs. Cory and Rip shuffled through the written reports.

"Look! Here is a report in Russian!" Cory exclaimed.

Rip yanked the paper out of his colleague's hand. "It's not Russian, Cory, you were just holding it upside down."

"No, I wasn't. It's Russian! Look at the—"

"It's not Russian, Cory, it's—"

"Wow, look at this crazy thing!" Ned interrupted the argument by holding up a picture of the Cactus Head.

Everyone looked at the black-and-white photograph.

"It's a decent picture," Sam remarked. "But it looks like it was taken at night without adequate light, so it is rather grainy."

Everyone agreed, but still they could easily see that the cactus-like image in the photo had arms and legs and appeared to be walking. "That's eerie," the twins thought.

Over by the maps, Melody spoke up. "The locations of the sightings are quite spread out. Ranger Douglas, where do you think is a good location for us to start our research?"

The ranger's eyes brightened at this question. "You guys aren't going to believe this, but there was a Cactus Head sighting just last night! So the location of that sighting will be a great place to start!"

The C.O.M. Crew members' mouths dropped open—what luck for them.

"Sometime in the middle of the night, three backpackers were scared away by the Cactus Head. They left all of their camping gear behind and got away with only their vehicle and the clothes they were wearing. They didn't get any pictures or video, but they did stop by the ranger's station to report the sighting."

A few minutes later, the Sassafras twins again found themselves in a car driving down the road in Joshua Tree Park. This time they rode with Symphony Douglas in her jeep. They drove down a small dirt road, leading the way as the six C.O.M. Crew members followed in the van. Blaine and Tracey had gotten a ride with the park ranger because when they had introduced themselves to her they had also expressed their interest in botany. She had cheerfully invited them to ride with her out to the abandoned campsite so she could chat about the local flora of the park on the way out.

She was just finishing a little speech about wildflowers. "So, there are many different varieties of wildflowers here in Joshua Tree and also in desert areas in general. They grow up and bloom in the spring and early summer months when it is warm and the grounds are relatively saturated with melting snow and recent rains. As the summer hits in earnest, though, the wildflowers die, and the seeds remain dormant until the next year."

The twins gazed out into the sunny national park as they bounced down the road, looking for signs of the wildflowers. They also began seeing more and more cactus the farther they drove. Especially a certain kind of rounded cactus that varied in height from pretty short to kind of tall.

"What kind of cactus are those?" Blaine asked, pointing to one.

"That is a barrel cactus," the ranger answered. "As you can see, it can grow about five feet tall, but it has a shallow root system. So, don't push these guys too hard, or they might fall over! Its characteristic cylindrical shape has multiple ridges. The ridges allow room for the plant to swell with water during the rainy season. The barrel cactus is extremely efficient at holding

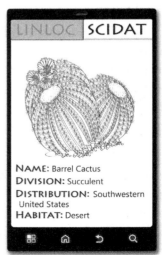

water, which is why Native Americans used it as a source of drinking water. They also stewed the inner pulp to make a cabbage-like dish."

Symphony took a breath before continuing. "The ridges of the barrel cactus also contain three- to four-inch spines. The Native Americans used these spines to make fish hooks because the spines of a barrel cactus are extremely tough and sharp. The spines can range in color from yellow to tan to red, depending on the exact species."

Symphony's jeep hit a huge bump in the road, causing all three to fly up out of their seats for a second, but that didn't stop the ranger from giving more information about the barrel cactus. "These cacti have more than spines, though; they also have flowers. The blooms don't appear every year, but if they do, the cactus will flower in late spring. It produces a ring of yellow-orange blooms at the top of the barrel. If pollinated, these flowers turn into a pineapple-shaped fruit, which animals love, but we humans find it a bit bitter. Another cool thing about the barrel cactus is that it will typically grow toward the south. The reason for this is to prevent sunburn. Because of this, they have been given the nickname 'compass cactus.' Isn't that amazing?"

"It is amazing," the twins agreed.

"Cacti, as a whole, are part of the succulent family of flowering plants. Most have a tough outer skin that holds water. The skin is usually covered with spines or spikes to protect the plant and prevent animals from eating it," Ranger Douglas said.

"So, do you think this Cactus Head creature is really half-cactus, half-human?" Tracey asked.

"Yeah, is it like a barrel cactus that has a face and grew arms and legs or something?" Blaine added.

"I don't know," the ranger responded. "There is just so much data that surely makes this creature seem real. It is more than myth, but still it is hard to believe, isn't it?"

The twins nodded.

"Well, hopefully, this C.O.M. Crew will help us find out once and for all if the Cactus Head is real or not. I'm excited to see what they can do."

The two vehicles kept driving out deeper and deeper into the park, and the twins didn't mind a bit. Joshua Tree was a beautiful place, and both Blaine and Tracey were awed by the breath-taking landscape. The plant life wasn't the only thing worth seeing. The rock formations they saw were also amazing. The twins had never seen anything like it. There were jagged rocks, smooth rocks, multi-colored rocks, and even rocks that looked as if huge marbles were stacked on top of each other. It made the Sassafrases feel like they had found a new planet.

"How far are we going to drive?" Blaine asked, enjoying the drive but wondering when they would arrive at their destination.

"The place we're going is out where the streets have no name," Ranger Douglas replied. "But we are almost there."

Just minutes after Blaine's question, they pulled up to the campsite that had been abandoned by the three backpackers. The ranger and the twins hopped out of their vehicle, followed by the mystery hunters climbing out of their van. The sun shone and the wind blew, but none of the nine spoke a word because the sight before them left them breathless and horrified.

The campsite looked completely destroyed. All sorts of camping gear and hiking equipment were strewn all over the ground. There was an untied boot, a slowly leaking canteen, a pair of pants with only one remaining leg, a flattened string of firecrackers, a smashed bullhorn, a blinking flashlight, and an inside-out sleeping bag on the rocks.

The three tents, though, were the eeriest looking of all. They had all been torn open by something apparently very sharp, leaving shredded material flapping hopelessly in the hot desert wind.

Whatever had attacked this place seemed to be wild and brutal. If it had indeed been the Cactus Head, the twins were second-guessing the decision to seek this creature out.

The rest of the group slowly shuffled in. Upon closer inspection, they found that there were numerous cactus spines stuck on and protruding from many of the disheveled items. Did this mean the Cactus Head had done this?

The twins were not sure, but the discovery of spines had sent the C.O.M. Crew into action. They ran back to the rainbow C.O.M. Carriage and started pulling out gear. They began setting up camp just a stone's throw away from the destroyed campsite in the same fashion as they had done in Redwood National Park. As they did, the eeriness of their setting began to fade, making way for the excitement of a new adventure.

Ranger Douglas and the Sassafras twins helped with the setup, and after the group had the tech station, a nice new fire pit, the perimeter sensors, and all the tents set up, the C.O.M. Cones came out. Ranger Douglas was delighted to see this new gadget, and unlike her cousin, she jumped at the chance to use one.

After a quick lunch, the group all equipped with cones, walked slowly out into the desert in search of the Cactus Head. Once again, they left Ned behind, but this time it was so he could get some sleep after driving through the night. Here in Joshua Tree National Park, there would be no fog or towering trees or bears; rather, there was sand-filled winds, rocky cliffs, and the possibility of dangerous desert animals like poisonous snakes. Those facts, coupled with the recent experience of being picked off one by one by the fake Bigfoot, caused everyone in the group to stick close together as they searched.

The sun shone, and the wind blew. The search took the whole afternoon, but it bore no results. No one saw a thing.

Around dinner time, Melody decided they should return to camp. They would all eat a good supper and wait until dark to resume the search.

On the way back, while climbing over a big mound of smooth rocks, not too far from their campsite, Blaine and Tracey got the chance to get a nice close-up picture of a barrel cactus. They stopped, took off their C.O.M. Cones, and pulled out their smartphones as the rest of the group trudged on ahead of them.

"What do you think about this whole Cactus Head thing, Trace?" Blaine asked his sister as they snapped pictures of the cool-looking cactus. "Do you think it really exists?"

"I don't know." Tracey shrugged her shoulders. "But to be honest, it does seem a little far-fetched to me."

"Yeah, you're probably right," Blaine agreed.

The twins set down their phones so they could put their C.O.M. Cones back on. Then, they stood up and hustled to catch up with the group. Cactus Head or no Cactus Head, they really didn't want to be alone in this desert. They caught up quickly, but what they didn't realize yet was that Tracey had accidentally left her phone lying on a rock in the middle of the Mojave Desert.

Chapter 17: Solving the Cactus Head Mystery

Creeping Creosote

The sun disappeared somewhere beyond the wide desert horizon, leaving the land lying in darkness. A light rain began to fall, floating in on the wind, but then ceased almost as quickly as it had begun, leaving in its wake dampness and eerie silence.

The seekers of mystery had full bellies from another fireside meal. They began to gather their gear and prepare to resume their search. Unbeknownst to any of them, something began to creep toward their campsite. It moved hauntingly yet forcefully like an invisible fog uttering no sound. Its movement seemed simultaneously quick and slow, as it got closer and closer successfully remaining undetected. Then, before any of them knew what had happened, it made its assault.

"Oh my," Tracey exclaimed. "What is that smell?"

"Oh c'mon, Rip," Cory said, scrunching up his nose. "If hot dogs treat your stomach this way, you just shouldn't eat them anymore."

"Hey! This smell is not my fault!" Rip defended. "It's—"

"Wow!" Blaine interjected. "It just kind of crept up and slapped us in the face, didn't it?"

"Wait, wait, wait you guys," Harmony interrupted. "Don't you remember what some of those reports said about a mysterious bad odor?"

"That's right, Harmony," Melody agreed. "A few of the reports said that a strange and weird smell proceeded an appearance of the Cactus Head."

Chapter 17: Solving the Cactus Head Mystery

Everyone became quiet and stared out into the darkness beyond their campsite. Was the Cactus Head out there right now watching them? Would it do to their campsite what it had done to the three backpackers? The tension among the group had suddenly become thick.

That tension, however, was soon cut with a chuckle from Symphony Douglas, "Everyone can rest easy," the ranger said with a good-hearted laugh. "It's a creosote bush."

"A creosote bush?"

"Yesiree, a creosote bush," Symphony replied. "Its leaves produce a pungent smell that is most notable after a quick summer rain."

"So, that's what this smell is?" Cory asked. "It's not the Cactus Head, but the leaves of a bush?"

"Yep! The leaves of a creosote bush have a waxy resin coating that helps to prevent water loss in the desert. This slightly stinky resin also protects the plant from being eaten by almost all mammals and insects. Native Americans used the leaves as medicine for pain and diarrhea, either by chewing them or by making tea."

The whole group nodded. Even though Blaine and Tracey were the only ones there studying botany, you could tell everyone was impressed by the information that Symphony was sharing.

"The creosote is a very slow-growing bush," the ranger went on giving more scientific data. "It can take up to ten years for the plant to reach just one foot in height. Even so, they can live for a very long time since they are well adapted to their environment. They are an evergreen shrub that is easy to identify because of the small yellow flowers that they produce in the late spring and summer. The flowers, which are usually less than one inch across, are bright yellow and contain five petals. Once they are pollinated, they develop a seed that rests inside a white wooly capsule. This seed can be carried by the wind, and many of them germinate during the rainy season. However, most of these new plants die out as the water dries up."

Tracey soaked up what Symphony had just said, and then mentioned thoughtfully, "I guess the desert can be a pretty harsh place for plants to survive."

"It sure can be," the park ranger replied. "The desert can have extremely hot days, and then at night temperatures can dip below freezing. Deserts cover about a quarter of the Earth's surface and, on average, receive less than ten inches of rain per year."

The group sat quiet for a moment dwelling on the desert. Then, Cory spoke up. "So, the weird smell that crept up on us is not the Cactus Head, but it's the leaves of a bush?"

"Oh my goodness, Cory, yes! You just asked the same

question a couple of seconds ago!" Rip retorted, arguing as usual with Cory. "There is no way the ranger could have explained it any better! It's—"

"Look!" Symphony interrupted Rip. "There is a creosote bush right at the edge of our campsite. Everyone grab a flashlight, and let's go take a peek at it."

The group did as the ranger suggested, and soon they were all standing over the short bush, illuminating it with their flashlights and holding their noses because of the stench.

Blaine and Tracey both reached for their smartphones so they could take a picture. Blaine got his, and snapped away, but Tracey couldn't find hers. She felt around in her pockets and looked through her backpack, but she found no phone.

"Oh no! This is bad!" Tracey's mind raced. "Where is my phone? I have to find it! What happened to it? Did the Man with No Eyebrows silently steal it? Did I accidentally leave it somewhe—"

All at once, she remembered she had left her phone out on the rocks by where she and Blaine had stopped to take a picture of the barrel cactus. Tracey ran out alone toward the rock where she had left her phone. She was so frantic that she didn't really take into consideration all the reasons that this course of action was not a very good one, nor did she explain to the others what she was doing.

Putting one foot in front of the other, she ran through the dark desert with the aid of a flashlight, thinking only about finding her phone. She hoped she could remember exactly where the spot was that she had left it. She recalled that it was a pretty big mound of smooth rocks, dotted with several barrel cacti. It wasn't too far from the campsite, but it sure seemed like it when you were looking for it in the dark. Sure, she had a flashlight, but its beam illuminated only a small circle in front of her and made the darkness outside of that small circle feel like walls. Nevertheless, she raced forward, through sand, and around plants, until she finally reached the smooth rocks. She pointed her flashlight down and then up, and saw that this was indeed not a single rock but a mound of rocks.

Tracey climbed up, trying to make sure that her feet and right hand used good spots on the rock that would aid her in going up. In her left hand, she held the metal flashlight, which she tried not to bang too hard against the rocks as she climbed. The Sassafras girl went up slowly yet consistently, and reached the top of the mound without slipping.

She waved her beam around, with hope in her heart, looking for her phone. The small circle of light swept left, and then right, and then every which way, but it showed no phone. Maybe if she could first find the barrel cactus that she and Blaine had taken a picture of, then she could find the phone. That is, if it was still there.

As she swept the beam of light around, she pinpointed the location of several barrel cacti, but figuring out exactly which one she had photographed was going to be hard. Tracey took a few careful steps to the closest one, and then looked around on the rocks—nothing. At the next cactus, there was no phone. The third cluster of cacti bore no results, either.

The most likely conclusion was that the Man with No Eyebrows had used the Dark Cape suit to steal her phone. The mysterious man had almost gotten Blaine's phone in France, and now it looked like he had finally prevailed and stolen a Sassafras smartphone. Tracey clenched her teeth. Who was this Man with No Eyebrows? She and Blaine had to find out, and they had to stop him. They had to get her phone back.

The Sassafras girl was frustrated. After just one more scientific topic, the glitch would have been fixed, but now with her phone gone, that wasn't going to be possible. Tracey flipped the light beam forward, splashing light haphazardly around on the top of the fruitless mound. Wait! What was that? Something had reflected a little bit of the light!

She focused her beam in that direction and . . . she saw it. Lying now only ten feet in front of her was her missing device. She bounded over to it, scooped it up in victory, and even gave it a little hug. Tracey knew that was silly, but she was just so relieved that she had found it.

The twelve-year-old put the retrieved phone safely in her backpack, gripped her flashlight, and prepared to climb down and head back to camp. Before she did, though, she swung the beam of light over to the barrel cactus that she and Blaine had photographed earlier. She mouthed a little "thank you" as if the plant had somehow aided in protecting her phone.

The girl now noticed something that she hadn't before. Another cactus was directly behind the one they had taken a picture of.

"I don't remember there being two cacti right here."

She shined the beam straight and steady, but suddenly, she saw something that made the beam shake. The second cactus had a face! Right then, as she was looking at it, the mouth of the face opened up and screamed, "Criers!"

The scream was so terrifying that it caused Tracey to drop the light. Clink, clank, clink went the metal flashlight down the mound of rocks with the light ricocheting in all directions until the battery-powered gadget hit the rock hard enough to click the light off, leaving the Sassafras girl all alone, in complete darkness, with the Cactus Head.

Tracey felt faint and fell to her knees. She was in jeopardy of falling all the way down and becoming unconscious, but before that happened, the Sassafras in her took charge. It was the never-give-up not-gonna-quit attitude that had sustained Sassafrases for centuries. Instead of falling down unconscious, Tracey bolted up off the rock and took off in the direction that she thought was away from the Cactus Head.

She ran forward cautiously yet quickly. She thought she was making progress until the sound of footsteps reached her ears and the shout, "Criers!" pierced the night once again. From the sound of it, the Cactus Head was even closer to her now than it had been when she had spotted it.

Now throwing every ounce of caution to the wind, Tracey sprinted into the darkness. She immediately paid for it. Her sprinting steps hit rock, rock, and then nothing. The Sassafras girl now felt herself careening helplessly into blackness.

"Where is your sister?" Ranger Douglas asked Blaine.

"She's right over . . . she's . . . I don't know."

The two looked together around the entire campsite, which wasn't terribly big, and Tracey was not there. They were about to inform the rest of the group about the girl's absence when suddenly a scream resounded from out in the desert. The ranger, the boy, and the C.O.M. Crew heard it clearly. The word that had been screamed was, "Criers!"

"It's the Cactus Head!" Rip yelled.

"And Tracey is out there alone with it!" Blaine yelped.

"She is?" Harmony asked, her face full of concern.

Blaine nodded.

"Everyone finish getting your C.O.M. Cones on," Melody instructed. "The search is on."

CHAPTER 17: SOLVING THE CACTUS HEAD MYSTERY

"Ouch." Tracey whimpered. Did she have any broken bones? Surely, after that fall she should have broken bones. She had come to rest on a large flat rock.

The twelve-year-old lay there thinking she might never move again, but then she heard footsteps and raspy labored breathing. These sounds spurred her up, and she soon felt herself running again. Evidently, no bones were broken, which was good, because if this Cactus Head thing was going to be relentless in pursuing her, she was going to need her bones intact.

She was off the flat rock in a flash. The moon and the stars were trying to shine, but there was quite a bit of cloud cover, leaving the desert-scape dark. Even so, Tracey's eyes had adjusted a little to the night so she could make out the shapes of most of the rocks when they were directly in front of her. Having a working flashlight would have been ideal, but she was managing now okay with just her eyes and the sparse dimness.

The Sassafras girl kept moving forward and soon found that the rocks she was traversing were changing from a mound to more of a stack or pile. For the most part, they were still rounded and secure, but now she had to deal with more gaps and holes. The new obstacles, however, were not stopping her. Tracey jumped, crawled, tunneled, and climbed over, under, and through the new rock formations with determined agility.

She wasn't sure how close the Cactus Head was to her now. There had been no recent shouts of "criers!" She hadn't heard its footsteps or breathing for a while, but that very well could be because the overwhelming sound in her ears was the beating of her own heart.

Tracey jumped up on a rock that was shaped like a big bowling ball. She then slid off it and went underneath the next circular boulder. She found herself in a short tunnel that went down and then back up. She burst out of the tunnel and now found herself running up an incline zigzagging around cone-shaped rocks.

Eventually, the rocks began to flatten out compelling Tracey to run faster.

Her zigzagging run morphed into a full-out sprint and then . . . she reached an edge . . . an edge for which she was running too fast. The Sassafras girl again found herself falling into darkness. Her body plummeted downward, but this time, she did not hit any rocks. Instead, she landed with a thud on a bed of sand. The landing had not broken any bones, but it had definitely knocked the air out of her lungs. Tracey sat up and moaned, gasping for air.

Tracey put her hands up on her head and was finally able to take some long deep breaths. She squinted through the dimness, trying to see exactly where she was. Now that she was on sand, she hoped that all the rock climbing was done and she could simply make her way through flat desert back to the campsite.

The girl got to her feet, put her hand on the rock wall next to her, and began making her way forward to see where the rock ended and the wide desert began. However, to Tracey's shock and dismay, the rock wall never ended. After walking in a circle once and then twice, the twelve-year-old realized that she had fallen into . . . a pit. It was a wide sandy pit, with huge rock walls on every side, and unfortunately, no part of the walls seemed easily climbable. Tracey knew that if she was going to get out, she was going to have to climb out, so the girl immediately began searching the rock that surrounded her for handholds and footholds or any place where she thought she could hoist herself up and out.

Tracey searched silently except for the racing thoughts in her mind. Then, in the silence, she heard something that threatened to turn all thoughts to despair. It was the sound of low heavy breathing. Tracey was not alone in the sandy pit. The Cactus Head was in here with her.

Protruding Paddle Cacti

"Criers!" came the loud screech again from the mysterious

spike-covered creature.

What did this call of "criers" even mean? It didn't make any sense, Tracey thought, but then again neither did the existence of a creature that was half-human half-cactus.

The girl's mind filled with a short list of possible explanations. The creature was truly a monster that was some sort of mixed freak of nature and science. The creature was just a figment of her imagination. The creature was just a human who was pulling a prank trying to scare her and everyone else. Maybe it was Ned again. No one had seen him since he had gone to take his supposed "nap" in his tent. Maybe he actually had the audacity to pull a scary prank again.

Tracey could barely make out the shadowy outline of the Cactus Head standing there in the middle of the pit. Whatever the explanation and whatever this thing truly was, it was scaring Tracey to the core. All at once, it began to move toward her.

Tracey shuffled quietly to her left, keeping her back against the stone wall. The creature saw her movement and pursued her in the direction that she had gone. Tracey streaked back to the right, hoping to throw the Cactus Head off her trail. It worked for a second, but then the creature adjusted, and followed her in that direction. Tracey shuddered. This pit now felt like an arena for this horrifying game of cat-and-mouse that she was engaged in.

The twelve-year-old darted this way and that, prohibiting the Cactus Head from catching her, but she didn't know how long she could keep this up. Surely, this mysterious creature would eventually find a way to get its spiky hands on her.

Then one of the worst things that could have happened, happened. Tracey tripped. She fell headlong into the sand, and when she managed to flip over on her back, she found the dark figure of the Cactus Head standing directly over her. Its mouth opened wide, and the now familiar scream echoed throughout the black pit and then up into the desert sky—"Criers!"

The creature stretched its arms out, and Tracey was sure that it was about to lunge out and get her, when suddenly the sandy pit was illuminated with light. It was like someone had reached out and flipped on an invisible light switch. The Sassafras girl could now clearly see this creature that was standing over her. Yes, it was basically covered in spines from head to toe, but it was not half-cactus, half-human. It looked to be all . . . human.

It was a man, not a plant or a monster. Tracey was pretty sure it wasn't Ned in a Cactus Head suit, either. She was about to say something to it when she was interrupted by a shout from somewhere beyond the light.

"Stop, Cactus Head!" It was the voice of Rip the mystery hunter. "Don't you move one inch closer to that girl! I have a tranquilizer dart with your name on it, and I'm itching to use it!"

"Tracey?" She heard Blaine's voice now. "Tracey, are you all right?"

Tracey looked up into the light and nodded. The girl now began hearing the familiar voices of the other C.O.M. Crew members from all around and above her, and she realized that the light was coming from the floodlights on their C.O.M. Cones as they stood in a circle around the rim of the pit.

The cactus-spine-covered man also looked up into the light, and Tracey saw there was no anger or malice on his face. Rather, his expression was one of anguish, pain, and helplessness.

"Criers," he said again, but this time not in a yell, but a pleading whimper.

"It's a man, not a cactus!" Cory yelped.

"And he needs our help," Harmony added with compassion in her voice. "Just look at him. Look at the pain he is in."

Within a few minutes, the crew had found a way down into the sandy pit at a place that was almost like a natural rock staircase. Tracey couldn't believe that she had missed the spot earlier. The

C.O.M. Crew, along with Blaine and Ranger Douglas, carefully led the bruised Tracey and the cactus man back to their campsite.

On the way, it was very apparent that Tracey was going to be just fine. They also realized that this mysterious man covered in cactus spines was not one to be feared. He posed no threat to the group. He was just a man who was in obvious pain and needed some help.

When they reached the campsite, the crew gave Tracey and the man bottles of water. Tracey gulped her bottle down fast and let everybody know she was okay. The Cactus Head mumbled the word "criers" as the Albermully sisters helped him sip the water.

Now that he was standing right in front of them and well illuminated, it was easy to see how he could be mistaken for a cactus. He did indeed have cactus spines sticking out from his shoulders all the way down to his ankles, his clothes and skin were crusted over with desert dust and sand, and protruding from the top of his head was a huge piece of cactus stuck very securely. Could it really be possible that this man had been roaming around the Mojave Desert like this for twenty years, being mistaken for the dangerous Cactus Head creature every time someone had spotted him?

The more water he drank and the less parched his lips and throat became, the clearer he spoke. "Criers" changed to "kiers" and then to "pi-ers." Eventually, the man finally clearly said the word "pliers."

"Of course!" Melody understood. "Pliers! He needs a pair of pliers!"

Sam ran over the van and rummaged around until he found a pair of pliers and then hurried back. Melody took them and carefully started pulling spines out of the poor man one by one. He winced every time one came out, but he was looking more and more human.

"That is a huge piece of paddle cactus that he has stuck there on the top of his head," Ranger Douglas commented.

"What's a paddle cactus?" Blaine asked.

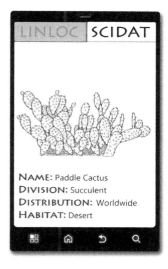

NAME: Paddle Cactus
DIVISION: Succulent
DISTRIBUTION: Worldwide
HABITAT: Desert

"Paddle cacti, which are also known as the prickly pear, look like they have 'leaves' that are paddle-shaped. These fleshy pads aren't really leaves but are actually modified stems that serve many functions for the cactus. They can store water, produce flowers, and perform photosynthesis. They are even edible once their spines are removed. There is also current research being done on the medicinal uses of the paddle cactus. All over the paddles are small dark wart-like projections called tubercles. Out of these tubercles come the spines of the plant."

As the female ranger spoke, Blaine took out his phone, and Tracey grabbed her retrieved phone. Both took zoomed-in pictures of the piece of paddle cactus that was stuck to the mystery man's head.

"The flowers of the paddle cactus can range from yellow to red to even purple." Ranger Douglas continued. "Once pollinated, they develop into a reddish-purple fruit that is delicious and sweet. The fruit is rich in nutrients and fiber so it is an excellent source of food. Fifteen species of paddle cacti are found in North American deserts and they can vary in height from one foot to seven feet."

As Symphony gave scientific information, the C.O.M. Crew worked together to try and clean the man up and ease his pain. Sam had found another pair of pliers, and he was helping Melody pull spines out. Ned, who had woken up and emerged from his tent, was helping the man drink as much cold water as he could get down. Harmony was wiping the dust and grime away from the man after the spines were removed. Cory and Rip were, of course, arguing.

"Lullaby, Cactus Head, lullaby to you," Cory was singing.

"Cory, he doesn't need a lullaby. He's a grown man, for goodness sake!"

"I just thought maybe a nice lullaby would bring him some comfort."

"If he was a cactus baby, maybe, but he is a cactus man. He doesn't want to hear your lullaby."

"Everybody likes a good lullaby when times are tough."

"No, they don't! It's—"

"How did he survive out in the desert like this for so long?" Blaine asked, interrupting Rip and Cory's argument.

Symphony just shook her head. "The Mojave Desert is home to many different species of cacti. All of which have developed specialized mechanisms to survive the heat and prevent water loss. Maybe in some strange way, this man was able to take on some of the traits of a cactus."

Suddenly, the night was pierced with yet another scream. This time, it was not the Cactus Head. It was Harmony Albermully, and what she screamed was half-question, half-exclamation.

"Michael?" the woman screamed loudly as she wiped the crust away from the cactus man's face. "Michael LaGrange?" Harmony repeated.

Everyone stopped what they were doing and looked at the man's half-cleaned face. A look of clarity was there now for the first time, replacing the look of pain and confusion.

"Harmony," he stated in a completely clear voice that sounded nothing like the raspy screeching screaming voice of the Cactus Head. "Harmony, my love, you found me."

Harmony's eyes filled with tears, and even though not all the cactus spines had been pulled out, she lunged forward and embraced the man.

"What? . . . Surely this wasn't . . . How could this man really be Michael LaGrange?" The twins' minds began asking presumably the same questions that everyone else was asking: "Was this truly the Albermully's long-lost high school friend? Why was he out roaming the desert? Why did he have a huge piece of cactus stuck to his head and spines all over?"

The next few minutes were a whirlwind of tears and laughter, hugs and gazes, pliers and winces, until finally Michael LaGrange sat before them spine-free and completely cleaned up except for the giant piece of paddle cactus protruding from the top of his head. Harmony held Michael's hand as Melody and Sam stood over him and each got a good grip on the hunk of cactus with their pliers.

Melody started a countdown. "Three . . . two . . . one!" Then she and Sam yanked up at the same time. Michael let out a small grunt as the piece of paddle cactus came cleanly off. Now the Cactus Head was no more. Everyone cheered, and then after a few more hugs and tears, they all sat down and listened to Michael LaGrange's story.

"Harmony . . . Melody," he said. "The last time I saw the two of you was the last day of class during our junior year of high school."

The Albermully sisters nodded, both teary-eyed, remembering the day well.

"The very next day, I set out with my family on a camping trip that my dad had been meticulously planning for months. My mom, my dad, my older brother, and I all piled into the family van and set out for Joshua Tree National Park. All of us were excited about the adventures that lay ahead. On the second day of our camping trip, my dad, my brother, and I decided to hike the tallest mountain of rocks that we could see from our campsite."

Michael paused here and looked reflective.

"It is amazing how clearly I am able to think without a big piece of cactus stuck in my head." Everyone smiled, glad that Michael was out of his misery and feeling more like himself.

"It was a fun climb, and the three of us reached the top of that mountain in no time. There is one thing you must know about LaGrange men; we are ultra-competitive. So upon reaching the top of the mountain, my dad came up with the idea for a little competition. He said that whoever ran down the mountain and reached our campsite first would be the winner for the day. Even though there was no prize or trophy for winning, both my brother and I, being as competitive as we are, were immediately game. My dad gave a countdown and shouted 'Go,' and the three of us were off.

"At first, we raced together in a tight pack, but then I looked off to my left and saw the opportunity for a short cut. My dad and older brother continued racing together neck and neck, as I shot off to the left on what I thought would be a much better path. At first, it was better. It was wider and less steep than the way that I had been going with my dad and brother, but before I knew it, it dramatically changed to a pathway of radical angles, loose rock, and lots and lots of cactus. I managed the trail okay for a couple of paces, and then the inevitable happened, and I fell. The last thing I remember clearly is careening toward a huge spiky cactus."

Michael stopped here and took a deep breath. He reached his hand up and gently felt the top of his cactus-free head. He smiled, took another deep breath, and then continued his story.

"I don't know how long I blacked out, but when I came to, I had no idea where I was. All I knew was, it was dark, I was outside, I was alone, and my head hurt. I reached up to feel the top of my head, and immediately recoiled in pain. I looked at my hand and saw that it was covered in spikes. I didn't know why. I was confused. My brain was foggy. I tried to move from the spot I was lying in, and as soon as I did, I felt more spikes poking me. I began to thrash around in a desperate attempt to get up, ignoring the pain and the fact that I was being continually poked. Eventually, I found my way to my feet. I looked and saw that I was covered in spikes, and for some reason, my response to seeing that was just to run out into the dark desert. I tried to call out for my family and my friends and for you, Harmony, but I couldn't seem to speak coherently. At that point, I am sure now, I began to lose my mind."

At this point in Michael's story, almost everyone was tearing up. It was heartbreaking to hear about all of the pain that LaGrange had had to live through.

"The hours turned into days, and the days turned into weeks and months and years. I kept roaming aimlessly through the desert, looking for shade in the daytime and sustenance at night. Clear thinking eluded me, but as I continually grabbed at the top of my head, one thought captured my brain—pliers. If I was going to get this prickly thing out of my scalp, I would need a pair of pliers. It was the only way to find relief. Somehow though, the word 'pliers' mixed with the anguish and loss in my heart turned into the word 'criers.' Every once in a while, I would see humans. I would hurry to them for help, but I only succeeded in scaring everyone away."

Michael paused again with a tear running down his cheek. "But tonight, twenty long years after my fateful fall. You all . . . the Creatures of Mystery crew . . . led by my dear friends Melody and Chorus and . . . the love of my life Harmony . . . you . . ."

The emotionally overwhelmed man couldn't finish his sentence, so Harmony finished it for him.

"We found you!"

Chapter 18: Back to the Basement

Surprise Jungle

Ahh to zip again! The countenances of both of the twins were bright as they zip lined through swirls of sonic light. They had only gone to one line location without using the invisible zip lines, but man, had they missed them! It was the Sassafras' opinion that traveling by invisible zip line at the speed of light was the most exhilarating experience in existence.

The location that they were zipping to now was their Uncle Cecil's basement, and they were excited about it. Not only because it meant that they would get to see their crazy uncle again, but also because it meant that they had successfully completed another scientific subject.

In the course of a week they had been to Peru, Scotland, Argentina, Borneo, Siberia, France, and both Northern and Southern California. They had nailed the subject of botany, just completely nailed it. Even with some precarious situations and that pesky Man with No Eyebrows working against them, they had remembered all of the topical facts, logged all the correct data, and learned about the plants found around the world. Botany was a subject that was now forever etched on their brains and hearts.

For some people, reading a book about science was good enough, but that's not how it worked for Blaine and Tracey. In order to remember and love science they needed to experience it face-to-face. They needed to interact with local experts and live through impactful experiences. They needed to be kidnapped by tribal warriors, framed as jewel thieves, sing soulful ballads, run from pirates, traverse speeding trains, race in sports cars, and hunt creatures of mystery. All these things helped to burn the information into their minds.

Chapter 18: Back to the Basement

At the California location they had just zipped away from, they had left behind a bunch of new friends and a completely redeemed man. How could they forget what they had learned about the paddle cactus when their basis of learning it had not come from a book but rather from the scalp of a mysterious creature?

The twins smiled as they thought about how Michael LaGrange had been reunited with his friends. After some phone calls and one more night sleeping in the desert, he had also been reunited with his parents and older brother. The smiles remained on the twelve-year-olds' faces even as they jerked to a stop when their light-speed zip-lining came to an end. The people they had met and the adventures they had experienced just seemed to keep those grins plastered there.

The customary post-zip-line tingling feeling prickled around the twin's bodies. They wondered if this was anything like what Michael LaGrange had felt when he was covered in cactus spines. His experience must have been much more painful.

The post-zip blindness and weakness also enveloped them, but these quickly dissipated, making way for normality. To the twin's shock, when all of their faculties returned, they saw that nothing about this basement landing was normal. As a matter of fact, they hadn't landed in their uncle's basement at all! They were lying on a floor of dirt and they were surrounded by all kinds of shrubs, bushes, flowers, and other plants.

What had happened? What had gone wrong? Maybe instead of being fixed, the glitch that President Lincoln had supposedly repaired had actually gotten worse. The twins were sure the coordinates that they had seen right before they zipped were the correct coordinates for Uncle Cecil's basement, but what was this place? This was not a basement.

Suddenly the Sassafras' heard the sound of something crashing through the jungly setting toward them. Something big and wild, that was not at all trying to approach with caution or stealth. Blaine and Tracey both hunkered down, their smiles long gone. The crashing got wilder and louder, and then bursting out of the greenery came . . . Uncle Cecil!

"Train! Blaisey!" The red headed scientist exclaimed with outstretched arms. "Welcome back to 1104 North Pecan Street!"

The twins were glad to see their uncle, but they had no idea what he was talking about. This was not his basement, but before they could ask him what he meant, they heard a smaller crashing sound. All at once President Lincoln, the prairie dog, jumped out from plant cover, but unlike Cecil who was dressed like normal in his messy white lab coat and bunny house slippers, the Prez was dressed to look like a miniature Incan tribal warrior. He had a tiny little wooden spear, a leather loincloth, and was even decorated in tribal paint. It was very reminiscent of what Blaine and Tracey had seen at their first botany location in Peru.

At this sight, the twelve-year-olds' smiles returned, but Cecil looked perplexed and Lincoln's frowned.

"Linc Dawg!" Cecil addressed his animal sidekick. "Why are you dressed like an Incan warrior? That is . . . Oh wait . . . now I remember . . . I am so sorry . . . I got so caught up in moving all the plants and dirt down here that I totally forgot. Sorry Linc!"

The prairie dog exhaled, obviously disappointed.

Blaine and Tracey had absolutely no idea what was going on. They voiced their confusion, "Uncle Cecil, where are we?"

"Why, you're home of course!" Cecil responded with a smile. "You completed your study of botany and you made it back to the basement!"

"But this isn't the basement." Blaine retorted.

"Oh good! I was hoping you would think that!" Cecil laughed.

"What do you mean?" Tracey asked.

"It was all President Lincoln's idea," the red headed scientist shared. "For your return home, after successful completion of your botany studies, the two of us would fill the basement with plants, so that the two of you would feel like you landed in the jungle again. It's a botanicalriffic celebration! We were also supposed to dress up like Incan warriors, but as you can see, only one of us remembered that part."

"You filled your basement with plants and dirt to celebrate us completing our botany leg of summer science learning?" Blaine asked.

Cecil nodded with smiling eyes.

"Awesome." Blaine chuckled.

Over in the basement at 1108 North Pecan Street, the floor

was not covered in dirt and there was no plant life to be seen, but there was something present that the eyebrowless owner of the house was very proud. It was about the size of a porta-potty, it was sitting directly in the center of the room, and he absolutely could not wait to use it.

All of his previous devious plans had failed. Even using the magic invisible Dark Cape suit had not worked, but now that this contraption sitting in the middle of his basement was finished and operational, he was brimming over with wicked joy.

This would indeed work in finally stopping those Sassafras twins from learning any more science.

He called it the Forget-O-Nator. He had spent the last few days putting the finishing touches on it. As usual, he had stolen most of the ideas for this invention from Cecil Sassafras.

A few years back, Cecil had the idea to make a machine that could remove memory and knowledge from a person's brain and place the memories in a canister. That way Cecil thought, he could literally share his thoughts to others. He started working on the invention, and had all the major components in place. Then, he had been confronted by that pet prairie dog of his, who evidently didn't think it was a very good idea to put the contents of your brain in a can.

Cecil eventually agreed, and the two had put the invention on the shelf. However, he and his eyebrowless-self had seen the parts, pieces, and plans in detail, through the lenses of his hidden cameras. He smiled and snickered now, as he looked at the finished product of one of Cecil's long forgotten projects.

The way it worked was simple—place the subject inside the Forget-O-Nator, lock the door, turn the one and only knob to full capacity, wait for two minutes, and open the door. Then remove the subject, whose memory would now be wiped absolutely clean.

His subjects, of course, would be those twins. He would

wipe their memories away, which of course included all the science that they had learned this summer.

He knew this meant that he would have to kidnap them somehow and bring them here to where the machine was. This would be easier said than done, but just like that knob on the Forget-O-Nator soon would be, his determination was cranked to full capacity.

He looked up at his monitors and saw that those twins had made it back to Cecil's basement, which meant they had successfully completed their study of botany. This time it did not even bother him because he knew that he would soon be able to wipe away everything that they had just learned.

Brutal confidence beamed from his eyes and a wicked laugh escaped from his throat.

"Revenge will come Cecil Sassafras," he rasped. "And it will come swiftly!"

Bonus Data

"So you didn't feel your phones vibrate?" Uncle Cecil asked.

"Nope, we sure didn't." Tracey answered. "I guess we just got so focused on our strange jungly landing that we didn't even notice whether our phones beeped or vibrated or anything."

"Well, go ahead and check them out!" Cecil encouraged giddily. "I bet there is something supertabulous waiting for you!"

Blaine and Tracey both knew what their uncle was referring to. Every time that they successfully completed a subject and made it back to the basement, their phones received a special text titled, "Bonus Data." Getting the information in this text was a sort of prize from President Lincoln and their uncle for their accomplishments.

Both twelve-year-olds pulled out their smartphones, and sure enough, there was some brand new bonus data waiting on the screens for them. Blaine jumped right in and began reading it aloud.

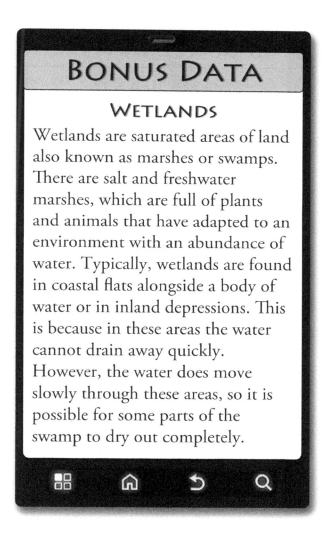

BONUS DATA

WETLANDS

Wetlands are saturated areas of land also known as marshes or swamps. There are salt and freshwater marshes, which are full of plants and animals that have adapted to an environment with an abundance of water. Typically, wetlands are found in coastal flats alongside a body of water or in inland depressions. This is because in these areas the water cannot drain away quickly. However, the water does move slowly through these areas, so it is possible for some parts of the swamp to dry out completely.

Cecil, as the giver of this bonus data gift, listened with bright eyes. Tracey picked up reading where her brother left off.

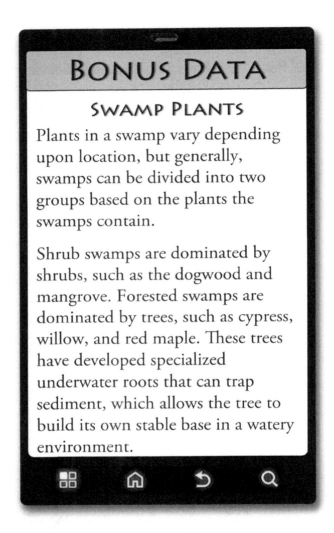

> **BONUS DATA**
>
> ### Swamp Plants
>
> Plants in a swamp vary depending upon location, but generally, swamps can be divided into two groups based on the plants the swamps contain.
>
> Shrub swamps are dominated by shrubs, such as the dogwood and mangrove. Forested swamps are dominated by trees, such as cypress, willow, and red maple. These trees have developed specialized underwater roots that can trap sediment, which allows the tree to build its own stable base in a watery environment.

President Lincoln, who was still dressed like an Incan warrior, stood there listening with bright eyes. Continuing the tag-team style reading, Blaine finished the rest of the bonus data.

BONUS DATA

Mangroves are typically found in wetlands along tropical coasts. Mangroves can tolerate the changes in the water level due to the tides. They are also extremely salt tolerant because of the ability to remove the excess salt they take in. The mangrove transports this salt from the roots to older leaves that will soon fall off or to specialized glands that excrete the chemical from the plant. Mangroves can have widespread arching roots that extend away from the plant. These prop-roots have the ability to "breathe" and take in additional oxygen for the plant.

Both twins exhaled in satisfaction.

"Thanks, Uncle Cecil and President Lincoln," Tracey said. "That was some great bonus data."

"And now to address the glitch, or lack thereof I might say." Cecil changed the subject on the fly. "Follow me through the jungle over to the computer and the tracking screen!"

Cecil and Lincoln excitedly turned and led the way, hacking their way through the fake jungle with the twins close behind.

President Lincoln reached the desired destination first, and proceeded to jump up on the computer desk, using his little wooden spear to tap around on the keyboard. The large screen on the wall behind the desk came to life, and soon everyone was looking at the tracking screen, which was the big map that continually traced the twins' movements as they zipped around the world. The screen usually had two green dots on it representing Blaine and Tracey, but right now, it was just a blank map.

"In a second, Linc Dawg is going to show the two of you something pretty coolio on the tracking screen." Cecil informed. "But first, let's talk about the glitch. As you know, the invisible zip lines, the LINLOC application, and the SCIDAT application are all linked. The zip lines transport you to and from different locations, LINLOC gives you the precise coordinates for each location, and SCIDAT allows you to record the data at the locations. But with the glitch, you two were prohibited from traveling freely. The glitch forced you to get all of the data entered exactly correct into SCIDAT. You could only move step-by-step through each location and couldn't just zip back here to the basement or anywhere else that suited your fancy. But now, after traveling to a line location without using the zip lines, and entering all your SCIDAT data correctly at that location, you have, according to President Lincoln, terminated the glitch!"

The prairie dog nodded and exuberantly lifted his spear above his head as an exclamation to Cecil's last sentence.

"On a little side note," the twin's uncle continued. "I know the lovely Summer Beach conversed with you two about other ways of possible travel via invisible zip line. One of which you, Blaine, got to experience first-hand."

An immediate cold sweat hit Blaine as he recalled what Cecil was talking about. Traveling the lines with randomly spun carabiner coordinates was no picnic. It had landed Blaine on a lonely island, a frozen lake, and nearly right into the heart of a fiery volcano.

"I think we can all agree that traveling the lines with random coordinates is too risky," Cecil said.

The twins nodded in agreement.

"The other way to travel is to use documented coordinates from the past and travel back to locations the two of you have already been to. So in theory, after the completion of zoology, you two could have zipped back to the basement or anywhere else you had already been at any time. You two were doing such a good job at moving forward that I honestly forgot to tell you that you could have zipped backward as well. Using a documented coordinate is, of course, how Summer was able to meet you in France posing under the alias of Été Plage."

"But it's also probably how the Man with No Eyebrows keeps popping up at our locations too, right?" Blaine interjected.

"Yeah Uncle Cecil," Tracey added with concern. "What are we going to do about the Man with No eyebrows? Who is he? How does he have access to our documented coordinates? Why is he stalking us and trying to stop us?"

Cecil Sassafras's perpetually joyful face sank for a brief moment.

"I have recorded all the times that this mysterious man has shown up at a location of yours and I have taken note of his villainy," the red headed scientist informed. "I will admit that this character is dangerous and seems to pose a major threat against the two of you prevailing in your summer-science-learning. I have no idea how he gained access to his own three-ringed carabiner and the invisible zip lines. I thought that the President and I had kept all of that tightly wrapped. I do have some ideas, however, as to who he may be, but there is currently no way to tell for sure. Regarding stopping him,

President Lincoln and I are working on a new application for your smartphones that will hopefully help you do just that, if he happens to pop up again."

The twins nodded again, excited about the prospect of a new app. They were a little concerned about the Man with No Eyebrows, but on the other hand the felt borderline invincible against the villain. He kept trying to sabotage them, but he always failed. They had successfully dodged everything he had thrown their way so far. It was not like he had some kind of machine that could stop them or make them forget everything they had learned. He was just a man. Yes, he was sneaky and tricky and relentless, but they felt he was also sloppy and very beatable.

"So. back to the glitch," Cecil effectively cut off the twelve-year-olds' thoughts about the Man with No Eyebrows. "As of now, there is no glitch! No glitch at all! Train and Blaisey, you can now move freely about the globe without the concern of perfectly entered SCIDAT info."

The Sassafras twins dwelled on this new fact for a second. Tracey wondered once more, as she had earlier when they faced the prospect of glitch-free zip-lining, does it really matter? They now had more freedom and less worry, but at this point, moving forward with science learning was the only direction that she wanted to go. Blaine felt the same way. Tracey was sure of it.

"Okay little man, hit it!" Cecil suddenly shouted to his prairie dog sidekick.

Lincoln stabbed his spear down onto the computer's keyboard, and the tracking screen immediately lit up with dots and lines of light that covered the image of the world map.

"President Lincoln and I just wanted to show you how impressive of a job the two of you have done learning science so far this summer," Cecil pointed at the map.

"The dots represent the locations that you landed at and sent data in from, and the lines represent the paths you took in between locations using the invisible zip lines. It is pretty amazeriffic how much science you have learned and how much of the world you have seen, is it not?"

It was amazing and terrific, the twins thought. Seeing it on a world map like this really did help put everything into perspective. There were dots on pretty much every continent. The lines representing their travels crisscrossed all over the place. Tracey was overwhelmed with a sense of gratitude and thankfulness that she had been afforded this amazing zip-lining, science-learning, and globe-trotting opportunity by her uncle. Blaine was wondering if it would be possible to get this image of the map printed on a T-shirt that he could wear to school in the fall.

"But you are not finished!" Cecil exclaimed. "Oh no nee-no-nee-no-nee-no! There are more lines to be traced across the world by the two of you! There is more science just sitting out there waiting to be discovered and experienced and learned! There is more adventure waiting as well, oh yes there is!"

Bursting with excitement and unable to control himself, Cecil attempted a spontaneous cartwheel. It went well until the landing. Instead of landing on his feet, he landed on his hind end in a patch of plants. He recovered quickly and bounced back up to his feet with one hundred percent of his excitement still intact.

"Train and Blaisey I know that the two of you are probably very tired from your botany adventures and are ready to rest, but if you will indulge President Lincoln and I for just a couple of moments, we have one more thing that we want to show you. It's about what is next!"

Blaine and Tracey knew they probably should be tired, but the word "next" had just electrified them.

"Give it to us!" Blaine replied. "Whatever you and Linc Dawg have for us, we will take it right now!"

Cecil had figured this is how his young relatives would respond. He was hopping over to another spot in the basement before Blaine had even finished his response. The joyful scientist shoved a few plants out of the way and revealed . . . something . . . Blaine and Tracey couldn't quite see it yet . . . Cecil moved a few more plants—it was . . . a sheet? Yes, a green bed sheet hung on the wall. Why on earth would their uncle want to show them this? But, instead of asking him what, why, or how, the twins remained silent and attentive and just waited to see what would shake out. Sometimes this was the best way to deal with their unpredictable uncle.

Cecil straightened the hanging sheet a bit and then used fancy hand motions to present it as if he were the host of an old-time game show.

"It's a green screen!" He exclaimed with smiles and laughter.

"A green screen?" Blaine and Tracey asked at the exact same time.

"Yippity skippity yesiree!" the scientist responded. "Linc Dawg, flip on the camera, and let's introduce these two to their next scientific subject!"

The versatile prairie dog smacked at the computer's keyboard yet again, effectively turning on a camera that was pointed at Cecil and the bedsheet. When the camera was turned on, all of the dots and lines of light that had been covering the world map over on the screen disappeared, and now what could be seen on the screen was the one and only Cecil Sassafras standing there looking somewhat like a weather man in front of a weather map.

The twins looked at their uncle right there in front of them standing by the green bed sheet, and then back over at him and the map on the screen.

"Ahhh, a green screen," they acknowledged with understanding.

Cecil slapped at his hair, attempting to comb it back a little. He buttoned up his lab coat, trying to make it look like a blazer. Then, he cleared his throat and began talking in his best news anchor voice.

"Good morning ladies and gentlemen. I hope you are feeling wonderful on what is shaping up to be a great day down in the basement at 1104 North Pecan Street. If I can, just let me jump right into what we can be expecting in the forecast for the next week or so. It looks like we have a one hundred percent chance of earth science moving in."

As Cecil spoke, he looked directly into the camera, and when he said "earth science." Lincoln hit a button, and the word appeared in very cool looking lettering on the screen. As Blaine and Tracey watched their uncle over on the screen standing in front of the map, talking professionally, and using weatherman type hand gestures, they really felt like they were watching a legitimate forecast.

"That's right ladies and gentlemen, we can definitely expect

loads and loads of earth science. And with that earth science there are sure to be some other fronts blowing in as well. We could get a sprinkling of coral reefs, and a concentration of rivers, oceans, and currents.

"There is also sure to be some climates and habitats heading our way," Cecil reported. "We can expect the study of weather, as well as tornados, hurricanes, and thunderstorms. There is a partly to mostly sunny chance that scientific natural cycles will develop, and we will also probably get a peek or two at the atmosphere.

"One dreaded thing to be on the lookout for, is the threat of a Man with No Eyebrows storm rolling in, but not to worry folks, because our young correspondents, Blaine and Tracey Sassafras, will be out in the field. The Man with No Eyebrows has proven to be no match for them. All in all, it's shaping up to be a beautiful week of earth science ahead, so press on, keep smiles on your faces, and never stop learning!"

Cecil finished his forecast and bowed to the cheers and claps of his niece, nephew, and pet prairie dog. Was the whole forecast thing a little cheesy? Yes thought the twins, but it was also fun and made them like their silly uncle even more.

They were excited about what was coming up next—earth science. It sounded from the forecast like it was going to be a very interesting subject to study. Also mentioned in the forecast, had been the Man with No Eyebrows. Any fear that the twins had over this mysterious man in the past, had pretty much faded into nothingness to make way for the determination they needed to beat him. Did they have questions? Yes, they had plenty of questions, but there was no real fear of him left in their hearts.

Back at the beginning of the summer, when they had bounced into Uncle Cecil's town on that bus, they had been haters of science. But now, after completing their study of three different subjects—zoology, anatomy, and botany, they were falling in love with science. Really, even though they viewed him as a villain, the Man with No

Eyebrows was partially to thank for that. All of his attacks, sabotage plans, and attempts to stop them had only succeeded in making Blaine and Tracey push forward harder.

They had turned from kids who despised all things science, to kids that liked science, to kids that loved science. Now, because of the continual threat of this eyebrowless man, Blaine and Tracey Sassafras were . . . defenders of science!

THE SASSAFRAS SCIENCE ADVENTURES

Stay in touch with the Sassafras Twins!

The adventure doesn't have to end just because you've finished the book! Connect with the twins and the other characters of the series through the Sassafras Science blog. You'll find articles in which:

- ✭ Uncle Cecil explains how to make a rocket at home;
- ✭ The Prez shares his wrap-up videos;
- ✭ Brock describes the Junior Ranger program;
- ✭ Summer details the process of photosynthesis.

Plus, Blaine and Tracey regularly pop in to say hi and share their thoughts. The Sassafras Science blog is ***the place*** to get to know the characters of the series!

The Sassafras twins would also love to keep in touch with you through their Facebook page. They share updates about future books, fun science-related activities, and cool nature news!

Visit SassafrasScience.com and click on the "Blog" tab to discover more!

Made in the USA
Middletown, DE
16 July 2021